Lecture Notes in Networks and Systems

Volume 77

Series Editor

Janusz Kacprzyk, Systems Research Institute, Polish Academy of Sciences,
Warsaw, Poland

Advisory Editors

Fernando Gomide, Department of Computer Engineering and Automation—DCA,
School of Electrical and Computer Engineering—FEEC, University of Campinas—
UNICAMP, São Paulo, Brazil
Okyay Kaynak, Department of Electrical and Electronic Engineering,
Bogazici University, Istanbul, Turkey
Derong Liu, Department of Electrical and Computer Engineering, University
of Illinois at Chicago, Chicago, USA; Institute of Automation, Chinese Academy
of Sciences, Beijing, China
Witold Pedrycz, Department of Electrical and Computer Engineering,
University of Alberta, Alberta, Canada; Systems Research Institute,
Polish Academy of Sciences, Warsaw, Poland
Marios M. Polycarpou, Department of Electrical and Computer Engineering,
KIOS Research Center for Intelligent Systems and Networks, University of Cyprus,
Nicosia, Cyprus
Imre J. Rudas, Óbuda University, Budapest, Hungary
Jun Wang, Department of Computer Science, City University of Hong Kong,
Kowloon, Hong Kong

The series "Lecture Notes in Networks and Systems" publishes the latest developments in Networks and Systems—quickly, informally and with high quality. Original research reported in proceedings and post-proceedings represents the core of LNNS.

Volumes published in LNNS embrace all aspects and subfields of, as well as new challenges in, Networks and Systems.

The series contains proceedings and edited volumes in systems and networks, spanning the areas of Cyber-Physical Systems, Autonomous Systems, Sensor Networks, Control Systems, Energy Systems, Automotive Systems, Biological Systems, Vehicular Networking and Connected Vehicles, Aerospace Systems, Automation, Manufacturing, Smart Grids, Nonlinear Systems, Power Systems, Robotics, Social Systems, Economic Systems and other. Of particular value to both the contributors and the readership are the short publication timeframe and the world-wide distribution and exposure which enable both a wide and rapid dissemination of research output.

The series covers the theory, applications, and perspectives on the state of the art and future developments relevant to systems and networks, decision making, control, complex processes and related areas, as embedded in the fields of interdisciplinary and applied sciences, engineering, computer science, physics, economics, social, and life sciences, as well as the paradigms and methodologies behind them.

** Indexing: The books of this series are submitted to ISI Proceedings, SCOPUS, Google Scholar and Springerlink **

More information about this series at http://www.springer.com/series/15179

Wynand Lambrechts · Saurabh Sinha

Last Mile Internet Access for Emerging Economies

Wynand Lambrechts
University of Johannesburg
Johannesburg, South Africa

Saurabh Sinha
Faculty of Engineering
and the Built Environment
University of Johannesburg
Johannesburg, South Africa

ISSN 2367-3370   ISSN 2367-3389   (electronic)
Lecture Notes in Networks and Systems
ISBN 978-3-030-20959-9   ISBN 978-3-030-20957-5   (eBook)
https://doi.org/10.1007/978-3-030-20957-5

This Springer imprint is published by the registered company Springer Nature Switzerland AG
The registered company address is: Gewerbestrasse 11, 6330 Cham, Switzerland

Preface

The last mile, with reference to the telecommunications industry and specifically Internet access, refers to the technologies that distribute connectivity from a point of presence to individuals, whether for personal or business use. The technology used for last mile access can either be wired, including copper cables and optical fibers, or wireless, including fixed wireless, light-fidelity (Li-Fi), through satellite networks, or traditional technologies such as wireless fidelity. The last mile is essentially a connection to the backbone of the network, also referred to as the middle mile, the technologies that bring the Internet to a central point capable of serving a community.

This book focuses on researching capable future technologies to deliver last mile Internet infrastructure in emerging markets and rural areas. The focus is placed on BRICS nations (Brazil, Russia, India, China, and South Africa) and the Global South relating to opportunities that enable these countries to drive economic development toward the fourth industrial revolution (Industry 4.0). These nations have additional limitations in establishing or expanding connectivity infrastructure that is not typically associated with developed nations. Limitations include lack of financial backing from local governments or private institutions, under-developed and unmaintained infrastructure, lack of skilled workers to oversee implementation and maintenance, unreliable energy from municipal grids, limited spectrum, high costs associated with purchasing end-user devices, and geographical restrictions in rural areas. The contribution of this book is to identify the limitations that do not necessarily apply to developed countries and categorize these to instantiate innovative policies to overcome the digital divide apparent in developing regions. An additional contribution of this book is presenting research into technologies that offer sustainable, long-term, upgradeable, and financially supported connectivity solutions in areas where traditional last mile connections are not practical. Contributory work emanating from this book includes research papers on scaling education in emerging markets to participate in Industry 4.0 and millimeter-wave integrated technologies toward Industry 4.0.

In the current information age, bringing connectivity to as many people as possible (ideally everyone) boosts socioeconomic benefits through innovation in science and technology, with a common goal of bringing positive change to the lives of individuals. Last mile Internet access in developing countries is not only intended to provide areas with stable, efficient, cost-effective, and effective broadband capabilities, but also to develop individuals living there to use connectivity efficiently for human capital development. This can be achieved by skills-training workers in areas that are geographically removed from high-quality primary, secondary, and tertiary educational facilities and providing them with opportunities to become self-sufficient in a sustainable way.

In addition, this book provides a strong theoretical background on not only future-generation technologies, but also current-generation solutions. The necessity to enlighten the reader about the basic principles of signal properties, which include signal propagation models, protocols, and hardware/software-related characteristics, is another focus of this book. To implement feasible future-generation technologies and ensure future-proof infrastructure that is easily expandable, it is of the utmost importance to understand the principles that influence the integrity of transmitted information. Maintaining last mile access is just as important as integrating initial broadband availability in emerging markets and rural areas. This book researches the importance of identifying, describing, and analyzing technology from a purely technological standpoint, but equally so acknowledges and investigates the sustainability of human capital through technology.

The primary audience of this book is learners in the fields of engineering and information technology who want to identify and act upon ways of advancing people in emerging markets and rural areas by bringing last mile Internet connectivity to individuals through future-proof technologies. The audience is presented with a theoretical background on signal propagation, basic networking, current last mile technologies, and in-depth analysis of next-generation technologies such as Li-Fi and millimeter-wave backhaul technology.

Acknowledging the Technical Peer Review Process The authors would like to recognize the support of the numerous technical reviewers, as well as language and graphics editors, who have all participated in the development of this research contribution. We value the system of scholarly peer review and the perspective that this adds to the production of research text that augments the body of scientific knowledge.

Johannesburg, South Africa   Wynand Lambrechts
                                                                                      Saurabh Sinha

Contents

Chapter 1
Bridging the Digital Divide: Innovative Last Mile Internet Connectivity Solutions

Abstract The global distribution of the internet varies greatly when comparing its accessibility and affordability in developed countries to that of emerging markets. The challenges that emerging markets face in distributing the internet to its citizens are increasing as technology evolve and infrastructures lag behind. Last mile solutions that are effective, feasible and maintainable are required to serve the millions of potential users that are currently disconnected in emerging markets. Economic growth has been linked to internet access and along with the incipient fourth industrial revolution, the need for broadband internet connectivity is also becoming increasingly important. Light fidelity (Li-Fi) and millimetre-wave (mm-wave) backhaul have been identified as potential last mile solutions in emerging markets, either as standalone technologies or as hybrid integrations with traditional solutions. This chapter researches the connectivity issues that emerging markets face and suggests possible methodologies to achieve last mile internet access that is specific to emerging markets.

1.1 Introduction—*The Last Mile* Internet Access

The last mile, with this book centering on specific reference to the telecommunications industry and therefore internet access, refers to the technologies that distribute internet and network connectivity from a point (the point of presence) to individual homes and businesses in a community. The technology used for the last mile can either be wired, including copper cables and optical fibers, or wireless, including fixed wireless, satellite or traditional technologies such as Wi-Fi. The last mile is essentially connected to the backbone of the network, also referred to as the middle mile, which are the technologies that bring the internet to a central point near the community.

In this book, the last mile is analyzed in context of developing countries, specifically the Global South. Rolling out the last mile in developed countries that have large financial backing and mature infrastructure has different limitations compared to that of developing countries. This book aims to not only identify these limitations, but proposes solutions to overcome them. The technologies that are available

© Springer Nature Switzerland AG 2019
W. Lambrechts and S. Sinha, *Last Mile Internet Access for Emerging Economies*,
Lecture Notes in Networks and Systems 77,
https://doi.org/10.1007/978-3-030-20957-5_1

to achieve the last mile are analyzed from a technical perspective, and additionally, evaluated based on various challenges associated with implementing them in developed countries.

The following paragraph summarizes the simplified distribution of the internet in order to identify and understand the significance of the last mile. Essentially, the internet is distributed into three primary categories, the core network, the middle mile and the last mile. Importantly, the complexity of each of these categories can be expanded significantly, as will be done in subsequent chapters of this book, but they are examined in their fundamental capabilities on the following paragraph.

1.1.1 The Distribution of the Internet

In a simplified representation, the internet is distributed in three primary categories, these categories are

- the core network (first mile),
- the middle mile, and
- the last mile.

The most commonly used middle- and last mile distribution technologies are copper cables and optical fibers, but these also include a variety of wireless technologies. The middle- and last mile must have reliable and high-bandwidth capabilities to ensure seamless and uninterrupted network connectivity to the core network. The infrastructure of both of these sections of the distribution network is of critical importance; however, its implementation is dependent on various external factors. Such factors include financial backing, skills-set of the technical support as well as geographical considerations. In Fig. 1.1, a representation of the infrastructure of internet/network distribution is shown, including the core network capabilities, the middle mile and the last mile towards end-users.

As shown in Fig. 1.1, internet and network distribution to individual homes or businesses in communities—both in indoor and outdoor environments—is typically divided into the three primary categories or sections as already identified in this paragraph. The first section, the core network, are technologies that route and authenticate network communication towards a service provider with the worldwide internet. These systems and technologies are typically powerful and exorbitant servers capable of handling large amount of traffic from numerous users. The core network is responsible for tasks such as aggregation, authentication, switching, service invocation and providing gateways for accessing other networks. Ideally, the number of core networks should be limited, with its data-handling capabilities optimized and with efficient distribution networks. This distribution is realized firstly by the middle mile infrastructure that aims to bring connectivity to the internet to a central point within a community, where further and broader distribution to end-users is possible. The technologies used in the middle mile should allow for high-bandwidth and low-loss transmission over relatively large distances, to ensure

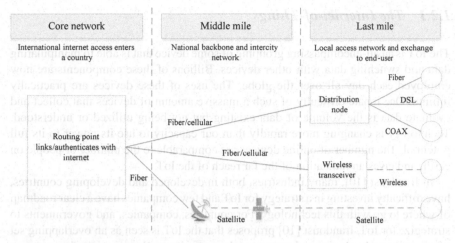

Fig. 1.1 A representation of the distribution of internet connectivity from the core network towards communities for broader distribution to individual users through the last mile

as little as possible signal degradation towards to the distribution point. Fiber links and satellite transmissions are commonly used in the middle mile as these are capable of handling high-bandwidth transmissions. Finally, the last mile distributes the network connectivity to the end-users through various technologies. These technologies should also ideally be fast and ensure little degradation of the signal, but the last mile also allows numerous technologies to be used based on the environmental limitations and end-user preferences. In addition, experimental technologies can be implemented in the last mile, with proper planning and redundancy, this is the ideal 'mile' to test and qualify novel distribution techniques.

In the following section, several key technologies, all integral in the fourth industrial revolution, are reviewed in terms of their significance for emerging markets to push last mile connectivity.

1.2   Why Push the Last Mile in Emerging Markets?

Several emerging technologies, all integral to Industry 4.0 development, have numerous advantages for emerging markets, and some of these are in their infancy, creating opportunities for these countries to spearhead the technologies and become central contenders in them. This section highlights several technologies that benefit from last mile access and can potentially boost the socioeconomic development of a country. The challenges and limitations associated with emerging markets are also highlighted in each paragraph. The first enabling technology reviewed is the internet-of-things (IoT).

1.2.1 The Internet-of-Things

The IoT is an all-encompassing grouping of some device that is able to manipulating data and switching data with other devices. Billions of these components are now employed each day all over the globe. The uses of these devices are practically infinite. One innate difficulty of such a massive amount of devices that collect and evaluate data is the volume of data existing but not being utilized or understood. Technology is changing more rapidly than our capacity to use its power to its full potential. The number of online devices are comparable to the number of people on earth, and gives an indication of the far reach of the IoT.

In Irandoust [10], many industries, both in developed and developing countries, have difficulty investing in a strategy for IoT and few companies have a clear roadmap of where to go with this technology. For countries, companies, and governments to strategize for IoT, Irandoust [10] proposes that the IoT is seen as an overlapping set of emerging technologies and sectors, and summarizes and categorizes these sectors as

- consumer IoT,
- enterprise and industrial IoT (IIoT),
- network and gateway,
- analytics, and
- autonomous.

In Miazi et al. [17], the opportunities and challenges of deploying IoT technologies in emerging markets are reviewed and presented. The findings in Miazi et al. [17] are significantly similar to the overall findings in this book in terms of last mile connectivity opportunities and challenges in emerging markets. It is therefore beneficial to briefly outline the key opportunities and challenges that were identified by Miazi et al. [17]. In terms of opportunities of the IoT for developing countries (IoT4D), with some overlap to the opportunities it presents in developed countries, are:

- In many emerging nations there is a broad deficiency of social security, and environments such as streets and areas are potentially unsafe, especially at night. IoT devices ranging from closed-circuit television to personal safety devices linked to a mobile carrier can be integrated into IoT platforms to provide an additional barrier of security for individuals.
- Accidents in the workplace is another common issue in parts where financial resources are limited and the correct health and safety procedures are not followed on a daily basis. IoT provides ubiquitous monitoring of various parameters within the workplace and can warn workers of threats such as fire, smoke, or toxic gases.
- Decentralized health management through IoT monitoring (such as heart rate and blood glucose levels) can overcome a lack of centralized hospitals and clinics in an area—especially in rural areas where it could take a fair amount of time to reach such an institution. IoT monitoring is capable of observing health statistics of patients remotely, and provide emergency services before fatal incidents occur.

- In developing countries, environmental monitoring is fairly common, as warning systems for natural disasters, or general monitoring of air quality. As technology advances and becomes cheaper, similar strategies using IoT devices should be implemented in emerging markets to safeguard the planet and avoid destruction of natural resources.
- Monitoring, analyzing, and pre-emptive policies on utilities such as drinking water, electricity, and gas through the IoT provide smart and cost-effective systems to minimize wastage and efficiently utilize these often scarce resources.
- Agriculture still remains an integral part of sustenance and sustainability of citizens in rural areas and in developed countries worldwide. The IoT has numerous advantages when employed in the agricultural sector, as evident in developed countries. Applications such as autonomous disease monitoring, nutrition level monitoring, and localized weather predictions can assist these citizens to yield higher crop production.

As with many technologies, the entire list of applications is practically infinite, and equally so with the IoT. However, developing countries are faced with different challenges to that of developed countries. In Miazi et al. [17] some of these challenges are highlighted, again traceable to the challenges of last mile integration in emerging markets. Some of the issues raised by Miazi et al. [17] are challenges in terms of low penetration of internet connectivity in developing countries and rural areas, unreliable energy sources, a lack of skilled workers, financial limitations, device reliability, and security and trust issues. These factors are also reviewed in this book, in lieu with last mile connections. There are however several successful initiatives of IoT in emerging countries and rural areas, to improve the quality of life of residents living in these areas. In Fernández et al. [4], some examples are highlighted, and listed in Table 1.1.

Countless IoT sensors by now monitor air quality, traffic, and water distribution in urban megacities in countries such as Moscow in Russia, Rio de Janeiro in Brazil, Beijing in China and New Delhi in India (all part of the BRICS (Brazil, Russia, India, China and South Africa) nations) [13]. Its uptake (which remains limited in rural areas) is also manifested in local governments that intents to intensify the effectiveness of data analysis and drive socioeconomic growth through enhanced resource management. The upsurge of IoT solutions, especially with respect to urbanized areas in emerging markets, is partly due to factors highlighted in Kreische et al. [13] which include

- technology prices have come down significantly,
- internet penetration in developing countries has grown,
- local governments have started to adapt policies to support IoT development after realizing its potential,
- remote collaborations through the internet have accelerated software development for IoT devices with specific requirements catered for regions in emerging markets.

Smart cities are also born of Industry 4.0 innovations and present several advantages from which emerging markets can benefit, reviewed in the following paragraph.

Table 1.1 IoT initiatives in emerging markets, adapted from [4]

Country	IoT initiative
India	Piramal Sarvajal has developed a low-cost reverse osmosis system coupled with smart controllers to efficiently distribute and monitor drinking water to certain rural areas in India
Indonesia	IoT water flow sensors and motion sensors assess the effectiveness of personal hygiene training delivered to parts of the rural population, for example monitoring if/how people wash their hands after using lavatories
Kenya (region of Kyuso)	The Oxford University has piloted a program that monitors the acceleration/movement of manual drinking water pumps through IoT sensors. Data are analyzed (in real-time at Oxford University) and potential failures at each pump can be detected
China (JiangSu province)	A variety of IoT sensors is used to monitor drinking water in key points in the distribution network
Bangladesh	A network of arsenic sensors has been implemented to monitor the quality of drinking water and prevent water pollution on a large scale
China/India/Africa	Pervasively distributed weather stations in many regions control irrigation pumps according to the current weather, saving water and improving crop yields

1.2.2 Smart Cities

This paragraph is adapted from Lambrechts and Sinha [14]. "A smart city uses currently available technologies to enhance living conditions through access to information about parameters that affect its inhabitants. These parameters include the status of its education and employment, utilities, transportation, energy consumption, health-related issues, water quality, air quality, waste management and any other relevant information that could potentially benefit the community. In a sense, many cities can already be considered smart, depending on the scope of technology and its reach. Reaching the status of a fully autonomous smart city is a long-term process and some cities aim to achieve such status faster, but there is no real quantity that qualifies a city as being smart, or not smart. The global market for striving for smart status is ever increasing with advancements in technology. Some areas of technology, such as miniaturization, have enabled implementation of sensors and actuators that have previously only been achievable in small quantities, as price and performance limitations could not be overcome. Challenges such as climate change, economic restructuring, online retail and entertainment, ageing populations, and public finances can be monitored and addressed using technology, and serve as a major drive towards such use" [14].

An urbanization trend witnessed in BRICS nations, is an imbalance between the distribution of urban dwellers and the growth in the number of cities [1]. The population density in existing cities have been attracting citizens from rural dwellings, and have relatively matured investment strategies and economic activities to manage the large, dense, populations. As smaller urban dwellings are promoted to cities, policy interventions are applied to limit growth, although these policies have not been successful [1], but many of these countries feel that the potential of the matured cities are underutilized. In Aijaz [1], numerous chapters are provided of smart city initiatives, policies, experience gained from successful pilot programs, finance, public engagement, security, and maintaining environmental sustainability in BRICS countries are presented. The South African Cities Network and University of Witwatersrand [21] provides a comparative and analytical overview of the urban developments across the BRICS nations, as well as a factsheet of 31 BRICS cities on general information and three primary themes: transportation, green energy, and innovation economies.

Industry 4.0 can thrive in a smart city, and in this book, last mile connections in rural areas in emerging markets are equally important as connecting urban environments and creating cities that adapt and analyze data in real time. As a result, the proposed technologies, policies, and strategies presented in this book are relevant primarily for sparse rural areas, but can be adapted and integrated in urbanized areas. Wireless sensor networks (WSNs) is an integral technology in smart cities, as well as sensing and monitoring conditions in rural areas, and typically require a consistent and reliable internet connection—provided through last mile connections.

1.2.3 Wireless Sensor Networks

This paragraph is adapted from Lambrechts and Sinha [14]. "A WSN consists of spatially distributed autonomous sensors to monitor physical environmental conditions and passes its data through (wireless) mesh network to a main location" [14]. "The general idea of a WSN is therefore a series of nodes, each node consisting of a variety of sensors, depending on the application and required information gathering. The vision of these networks is based on the strength-in-numbers principle, since a multitude of sensors (depending on the cost of each sensor) work together to provide accurate information on each location in a system. Community sensing mechanisms based on incentives could potentially provide a good starting point to encourage citizens to participate in information gathering [3]. Typically, several main subsystems make up a node, consisting of a transceiver and antenna combination, electronic circuitry that interfaces with the sensors and energy source, generally a power source in the form of a battery to provide mobile implementation, or an embedded form of energy-harvesting (such as solar power) to provide long-term and sustainable operation of these components"—especially relevant in unreliable energy from local utilities in developing countries. WSN have a great role to play in developing countries, capable of expediting novel solutions to mitigate development issues and facilitate research [26] in crucial scientific areas such as

- environmental monitoring,
- energy management,

Table 1.2 Challenges of producing and maintaining realistic WSN applications that are amplified in emerging markets, adapted from [22]

Challenge	Intensification in emerging markets
Making sense of large amounts of data (big data)	Lack of skilled workers. Lack of computing resources. Outdated algorithms
Reliable and robust systems	Financial limitations leading to low-cost (unreliable) electronic components. Lack of maintenance resulting in lack of skilled technicians
Security	WSN's wireless nature open to security attacks and data breaches. Outdated security mechanisms in non-maintained systems
Privacy	Ubiquitous nature of WSN leads to privacy concerns. Lack of political intervention and policies in emerging markets can lead to little accountability for wrongdoers
Constant and real time operation	Unreliable energy from gird. Lack of maintenance. Low-cost and unreliable components

- traffic control [8],
- agricultural practices, and in
- health and safety.

Challenges of producing and maintaining realistic WSN applications, amplified for emerging markets, adapted and presented in Table 1.2 [22].

Development countries with amplified levels of inequality, poverty, illiteracy, and a general lack of infrastructure [26] require additional research and development programs, and collaborative efforts to implement successful WSN application and maintain them, ensuring longevity and feasibility of such projects. The thematic approach of this book to provide efficient and future-proof last mile solutions aids in the tailored approach and longevity of WSNs in emerging markets. Another example of how technology is beneficial in automated structures and programs in various application, is geographic information systems (GIS), briefly reviewed in the following paragraph.

1.2.4 Geographic Information Systems (GIS)

This paragraph is adapted from Lambrechts and Sinha [14]. "GIS can be traced back to as early as 1854 when cholera hit the city of London, England. British physician John Snow began mapping outbreak locations, roads, property boundaries, and water lines during this period of the outbreak. He noticed that cholera cases were commonly found along the water line" [14]. "This could be considered the first time that data was gathered and analyzed, and visually displayed on a geographical map to realize a trend of a present health risk, in some ways, a classical GIS. GIS has evolved as a computer-based (to handle vast amounts of data without requiring a person to perform these sometimes tedious calculations) tool that analyzes, stores, manipulates and visualizes geographic information on a map. GIS requires skilled individuals from various disciplines to effectively display data in an organized and easy to use way, these disciplines include cartographers, database managers, programmers, remote sensing analysts, spatial analysts, and land surveyors. GIS is used in many professions, including agriculture, archaeology, architecture, business, education, engineering, environmental studies, consumer science, forestry, recreation, media, law, medicine, military sciences, public administration, public policy, real estate, social work, transportation, and water services. These systems are not only reserved for wealthy developed nations, and could benefit a larger audience globally, as it already has been doing for several years. Obtaining the goal of sustainable development within Africa and other developing countries' diverse communities requires that analysts and decision makers understand the characteristics of resource use as well as human conditions.

Maps are a fundamental part of GIS portraying collections of spatial geographical information as thematic (themed/relevant) layers and the art of map construction is called cartography. Types of maps range from topographic maps that show a variety of structures such as roads, land-use classification, elevation, rivers, political boundaries, and the identification of buildings. Specialized maps such as weather maps indicating low and high pressure systems which can become complex since the weather information and the map type must be precisely layered to convey accurate predictions.

GIS maps are used for communication and understanding large amounts of data in an organized way such as finding patterns, deriving new information using analysis, getting applicable status reports, compiling geographic information, communicating ideas, concepts, plans, and designs, and sharing geographic knowledge openly. Information conveyed on a map must be carefully transferred to avoid loss of information. The loss of information is however inevitable, as can be expected for example when transferring from a topographic map that abstracts three dimensional structures at a reduced scale to a two-dimensional map on a plane of paper. A technique to ensure accurate representations of the information through layer mapping is called translation or projection.

The key attributes of a GIS are the integration of geometric and thematic attributes of spatial objects. Geometric data describe the location of an object and thematic data describe the application-specific information of the object, generally gathered by sensors, surveys, or any other manual means of data gathering. GIS data capturing refers to the process of entering data into a GIS, generally in digital format as analogue to digital conversion was already applied by the time the data are entered into the system. Images taken by satellites and data tables can be used, or maps can be scanned in and uploaded to the system. The GIS takes the information from these sources and aligns the data according to scale to ensure it fits together in a single map. Additionally, a GIS must manipulate the projections of a map, a process to transfer information from the curved surface of the earth to a flat surface, which can result in minor distortion of the three-dimensional data. As a rule, any map can only show either the correct size of a land area or the correct shape, not both simultaneously."

Mennecke and West [16] reviews issues in data collection, implementation and management of GIS in developing countries. As stated in Mennecke and West [16], the role of centralized planning, management, and decision making is increasing in its importance in emerging markets due to pressure of

- overpopulation,
- exhaustion of natural assets and resources, and
- monetary volatility.

Spatial data gathered from GIS is critical in preparation and development strategies since they designate and map the dispersal of economic resources, population, and other significant elements [16]. Again, numerous additional challenges in developing countries such as

- difficulty in obtaining accurate and trustworthy data,
- incorrect handling of spatial data,
- political gain from manipulating or withholding spatial data, and
- issues in obtaining and keeping skilled workers

are plaguing GIS implementations, as with many other technologies. Finally, strategizing and tailoring policies and technologies presented in this book to deploy last mile connections to emerging markets and rural areas would be beneficial to overcome challenges and limitations presented in each of these categories, and to promote the beneficial characteristics of each.

A summary of the potential last mile technology drivers is presented in the following paragraph. These technologies are constantly evolving and at the time of writing, are considered the primary contenders to deliver last mile access considering that each distribution infrastructure is different based on the geographical context and number of users it should be able to provide for, along with their respective bandwidth necessities.

1.3 Technologies that Can Drive the Last Mile

The last mile connectivity can also take advantage of various types of technology to not only ensure redundancy, but also allow for alternatives that have specific advantages based on the physical environment in which they are employed to ultimately provide broadband access. Five major technologies that are traditionally used to deliver broadband internet access to end users are

- mobile connections on Global System for Mobile Communication (GSM) networks—with 5th generation (5G) mobile connection the driving force of modern connectivity,
- digital subscriber lines (DSL), cable modems (copper-based) or power line communications (PLC),

- optical fiber,
- fixed wireless, and
- satellite communications.

In additional to these five major technologies, newer technologies such as visible light wave and infrared (IR) light wave communication are receiving added attention due to its high-bandwidth capabilities and abilities to be used in for example rural areas. Technologies such as light fidelity (Li-Fi) offer various advantages to achieve connectivity of the last mile. Each of these alternative methods of achieving the last mile has their own distinct set of advantages and disadvantages, depending on the proposed application, physical environment and political incentive to drive a specific technology.

This book looks at the detailed characteristics and technological aspects of each solution and furthermore proposes various infrastructure and development opportunities to combine technologies to deliver the last mile to communities that traditionally struggle to gain access to the internet, or have no access whatsoever. A brief summary of the main advantages and disadvantages of each of these five major technologies and Li-Fi is provided in Table 1.3, serving as a reference to the baseline technologies in subsequent reviews of the proposed infrastructures to achieve the last mile discussed in this book.

As seen from Table 1.3, several alternative technologies can provide broadband internet to virtually any environment, if planned for and executed appropriately. Each of these technologies present its own distinct advantages and disadvantages. The rapid growth of the digital economy is driving a clear shift in the evolution of broadband services for public and private sector use. To be considered an economically viable technology used by the last mile, the connection to the end users should

- deliver adequate amounts of signal power to the end-user to ensure uninterrupted access,
- not allow high levels of signal loss and degradation during transmission,
- enable support for large bandwidth/low latency communication to accommodate multiple users and/or broadband internet access,
- provide transmission signals with low noise compared to the signal intensity, therefore a high signal-to-noise ratio,
- provide access to the network at various locations that are easily accessible and reliable,
- integrate high levels of security to protect user's privacy, and
- be affordable to communities with respect to maintaining the core network, middle mile as well as individual user access.

These requirements of the last mile can become relatively complex to achieve, especially considering the longevity and sustainability of the connections. As a result, the workers employed to realize, implement and maintain the last mile should be trained and their skills developed continuously to warrant the viability of the employing such services. This is essentially one of the major challenges in the developing world, where basic education is often lackluster and not considered a high priority.

Table 1.3 The advantages and disadvantages of the five major technologies and Li-Fi capable of delivering the last mile to end-users in communities

Mobile (GSM) networks—advantages	Mobile (GSM)—disadvantages
High level of network access mobility	Broadband network access is costly
Relatively good service reliability	Cellular network access in rural areas are limited
Cost-effective to gain basic access to networks	Network speed varies with external factors such as weather and foliage
DSL/cable—advantages	DSL/cable—disadvantages
High level of network availability—especially in urban areas	Performance degradation a strong function of distance to the core network
Low infrastructure cost for access to homes and businesses	Speed limitations hinders future-proofing infrastructures
High service reliability and quick turnaround times for network maintenance and repairs	Copper cables prone to long-term damage and corrosion
Fiber—advantages	Fiber—disadvantages
Very high bandwidth capabilities at low latency	High cost for individual users
No susceptibility to electromagnetic interference	Glass wire cables prone to physical damage
Lightweight cables, service reliability and long lifespan	
Fixed wireless—advantages	Fixed wireless—disadvantages
Cost-effective to gain basic access to networks	Network speed varies with external factors such as weather and foliage
Standard protocols such as WiMax already established	Network reliability and transmission speeds dependent on line-of-sight access
Potentially deliver transfer speeds of several 10's of Mbps	High latency typically associated with fixed wireless
Satellite—advantages	Satellite—disadvantages
Accessible from virtually anywhere in the world	Network speed varies with external factors such as weather and foliage
High service reliability	High latency typically associated with connections
Li-Fi—advantages	Li-Fi—disadvantages
Low investment cost per household or business	Network reliability and transmission speeds dependent on line-of-sight access
Very high bandwidth transmissions	Always-on lights required for network access
Multiple users on single connection without degradation	Susceptible to environmental interference from other/ambient lights

Governmental policies and socioeconomic benefits of attaining such skilled individuals should be considered one of the highest priorities for such a country to succeed in providing internet to communities, something being considered a basic human right in modern times. The following paragraph highlights the fact that internet connectivity is considered an unofficial basic human right with reference to the United Nations (UN) Sustainable Development Goals (SDGs).

1.4 Internet Access—A Human Right

In modern times, internet connectivity is considered a basic human right; proclaimed by the UN in 2016, and all citizens are encouraged to be a part of the digital economy. According to the UN SDGs of the 2030 Agenda for Sustainable Development, investments in infrastructure of information and communication technology (goal #9 of 17) are crucial in achieving sustainable development and empowering communities towards achieving prosperity from the fourth industrial revolution. According to the SDG, three primary driving forces must be addressed to foster economic growth, these are

- infrastructure,
- industrialization, and
- innovation.

To support sustainable development, inclusivity, resilience and sustainability should also be factored into the implementation of these three driving forces. Furthermore, equitable quality education, specifically in developing countries such as sub-Saharan Africa and Southeast Asia can be achieved by providing communities with proper access to internet connectivity. Countries that have declared the internet as being a basic human right, already include

- France,
- Finland,
- Estonia,
- Greece, and
- Cost Rica.

It is however still a problem that persists, that internet connectivity services remain inaccessible across large areas of the developing world—making the gesture of declaring it a basic human right seemingly irrelevant if nothing is done to provide it to its citizens. It requires a commitment from a country to provide its citizens with internet access, even in the most remote areas, and many developing countries are realizing this and rolling out policies and campaigns to drive and incentivize this.

1.5   Developing Countries' Commitment to the Last Mile

As an example, in Johannesburg, South Africa, the Metropolitan Trading Company (MTC) drives an open-access fiber distribution network with a primary objective to bridge the digital divide commonly seen in developing countries and transform this city into a world-class African city [18]. MTC is responsible for connecting and maintaining broadband distribution across the city of Johannesburg, with local government adopting a strategic, creative and integrated approach to local governance in addressing challenges. Key strategic objectives of the MTC are

- to decrease the cost of communication to the end-user,
- serve as a facilitator in ensuring better service delivery,
- bridging the digital divide, and
- encouraging societal and economic growth in Johannesburg, South Africa.

The values of the MTC are also especially noteworthy and important, especially when considering that many developing countries are plagued with corruption and often with misplaced policies. The policies and values outlined by the MTC can therefore be used and applied to other projects and initiatives that aim to achieve similar goals. The MTC approaches its value system by focusing on, adapted from MTC [18],

- the customer—service and quality being primarily customer-focused,
- ethical conduct—dealings with subcontractors that share ethical conduct,
- integrity when making decisions that affect shareholders,
- efficiency in various sectors, including energy, materials and time, and
- innovation not only in current solutions but that addresses sustainability for future generations.

Furthermore, as outlined by the MTC annual report and commonly seen in developing countries, the challenges of broadband in developing nations and particularly in rural communities are the lack of constant, high-bandwidth and high-quality internet-enabled connections that is becoming increasingly important in public institutions, businesses, and for personal/private use and this negative affects the development of a country and its global competitiveness. Slow development within information and communications technology (ICT) and high costs associated to end-user access has led to mobile broadband becoming the primary form of access in these communities, a costly and inconsistent alternative—something not evident in mature economies [18].

MTC positions itself as a core metro link within the broadband industry and can be utilized to provide a series of services and products to the last mile providers. This positioning is crucial for MTC to impact the industry and lower the cost of communication for the residents of Johannesburg, and in future, for more communities within South Africa. Figure 1.2 represents the broadband industry and the positioning of MTC as a core metro segment.

A purely market-driven model excludes populations in rural developments of sufficient access to broadband internet, primarily driven by inflated fees and overly

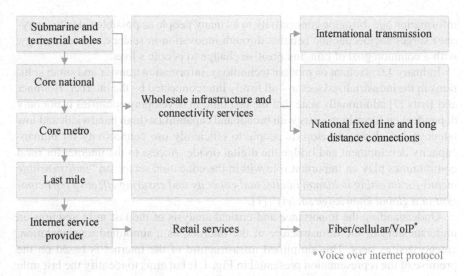

Fig. 1.2 A representation of the broadband industry and the segmentation of MTC, adapted from MTC [18]

complex policies that are not backed by local governments. As a result, new technologies as well as updated policies are being developed to provide basic internet access to communities that have historically been at the center of the digital divide. Public connectivity deployments are typically a combination of commercial internet hotspots and state-owned sponsored programs and can be beneficial for end-users if policies are set up to protect all parties involved. The following paragraph outlines the focus of this book with respect to connectivity for citizens that have been excluded due to various limitations and restrictions.

1.6 The Focus of This Book

This is also the focus of this book—taking advantage of the last mile to distribute internet connectivity to end-users, taking into account the countless limitations that may exist due to environmental conditions, living conditions, user requirements, geographical limitations and political precedence. Another specific focus of this book is applying the last mile in developing countries and rural areas, where the limitations are vastly different compared to urbanized areas and especially in developed countries. This book aims to distinguish the various limitations in rural and developing areas compared to that of developed nations. By categorizing these limitations, the technologies and policies that could enable internet access to users in such areas can be identified, explored, reviewed and proposed. Selective focus is also placed on BRICS nations and opportunities to enable these countries to drive economic development towards the fourth industrial revolution (Industry 4.0). In the current information age, bringing connectivity to as many people as possible (ideally every-one) surges socioeconomic benefits through innovation in science and technology, with a common goal of bringing positive change to people's lives.

Industry 4.0 is reliant on modern technology, information transfer and smart solu-tions in the industrialized sector—all firmly interconnected by the internet. Holmner and Britz [7] additionally state that the last mile in developing countries is not only dependent on furnishing areas with broadband capabilities (high bandwidth and low latency), but also to develop its people to efficiently use connectivity for human-capacity development and bridge the digital divide. Access to the internet for rural communities play an important role within the education sector, an *"indispensable means for investing in human intellectual capacity and ensuring effective participa-tion in a global knowledge society"* [7].

Understanding the importance and critical analysis of the last mile, requires an understanding of the infrastructure of the internet, as a simplified representation, categorized as tiers. This simplified infrastructure of the internet is based on the premise of the representation presented in Fig. 1.1, but aims to identify the last mile in the scope of the global distribution of the internet. Figure 1.3 represents the internet infrastructure as three primary tiers, where each tier embodies a substructure of the internet.

From Fig. 1.3, the first tier (Tier 1) of the internet infrastructure represents the internet service providers (ISPs) that connect large internet exchange points that typ-ically span across continents. These providers exchanges internet traffic with other Tier 1 providers typically through peering agreements. If an ISP is said to be peering, it opens their network to other Tier 1 providers without purchasing internet protocol (IP) transit or paying for this peering. This tier is essentially the backbone of the internet and the providers build infrastructures that allow for the exchange of the

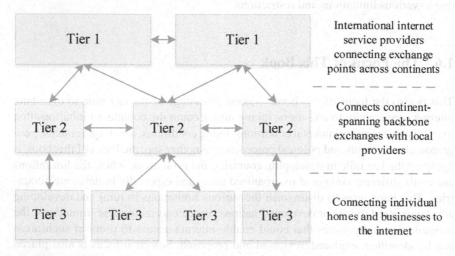

Fig. 1.3 A simplified representation of the three primary tiers of the internet infrastructure

internet among continents and countries. The second tier, Tier 2 as shown in Fig. 1.3, are tasked with connecting the continent-spanning backbone exchanges with local providers, typically through either peering agreements and purchasing IP transit. Tier 2 providers can then exchange internet connectivity and bandwidth allocation to the lower tier providers, Tier 3 as shown in Fig. 1.3. Tier 3 providers connect individual homes and businesses to the internet, and importantly, this is where the last mile of connectivity lies. The last mile is therefore of vital concern with respect to network performance for end-users. A significant portion of network degradation appears in the last mile and this can especially affect rural communities. It is therefore up to the Tier 3—last mile—providers to build, maintain and expand its internet infrastructure as requirements of network performance, bandwidth allocation and internet connections vary. In many developing countries, poor internet exchange points (Tier 2) and poor last mile infrastructure often results in internet traffic having to be diverted to other countries (or continents) before it reaches the user, significantly degrading the connection and hence the expandability and sustainability of the internet connection.

The last mile is not only applicable in telecommunication, but also has been used in challenging situations of extending water, sewage and electrification to end-users [2]. In this book, however, the last mile exclusively refers to the innovations, challenges and proposed solutions of delivering the last mile, with specific focus on rural areas and in developing countries. The significance of the last mile connectivity will also be discussed in respect to improving the educational system, enabling communities to take part in Industry 4.0 and elevating the quality of life in rural communities. In developed nations, the last mile is typically a large function of planning, efficiency and execution logistics; however, in developing countries, several additional challenges exist, including

• geographical limitations,
• lackluster or unsupportive governmental policies,
• shortcomings in funding and/or little interest from public and private investors,
• skilled workers to implement and maintain the infrastructure,
• sustainability for current and future generations,
• training of people to use the infrastructure, and
• maintaining the infrastructure technically.

These additional limitations are also discussed in this book, where proposed solutions are presented for both the technical and non-technical limitations. In order to initiate proposed solutions to aim to rectify the current situation in many developing countries, several steps should be taken, as outlined in Bull and Garofalo [2]. These steps include

• the current situation and state of the limitations and shortcomings must be recognized by schools and tertiary educational institutions to properly train new generations of the workforce to understand the problems, possible solutions and methods to maintain it,
• these institutions must also devise and communicate proposed strategies, policies and solutions to the upcoming generation of the workforce and be able to entrust them with taking it forward both in the short- and long term, and

- gains in learning and other achievements as a result of technology must be warranted by developing leadership policies since lack of leadership often results in undermining the benefits of access to technology [2].

Importantly, before being able to implement policies that enable communities to have access to for example stable and constant internet connectivity, technologies that enable last mile connections should be identified. These technologies should be

- accessible to a variety and large number of users and devices,
- cost-effective,
- able to provide high-bandwidth capacity to numerous users,
- sustainable and easily maintained,
- upgradeable,
- able to overcome geographical limitations (wireless vs. wired), and
- financially supported therefore be considered mature.

The following sections identifies various technologies that can be used to achieve the last mile in telecommunications, specifically in rural areas in developing countries. These technologies include traditional methods such as copper-based and fiber-based wired technologies, traditional wireless technologies, wired alternatives such as PLC and also technologies that are still maturing such as light-fidelity (Li-Fi) and millimeter-wave backhaul. The sections identify and describes these lesser-mature technologies, whereas the more traditional technologies are briefly discussed in subsequent chapters of this book. The reason being to enlighten the reader of these technologies and also to analyze its viability in environments where traditional technologies have not yet been able to penetrate in serving large communities in developing areas. It is also important to note that all of these technologies could be used in unison to provide solutions for a variety of scenarios, therefore they are not mutually exclusive. The first section introduces light fidelity technology as a viable alternative to implement the last mile.

1.7   Light Fidelity (Li-Fi) Technology

Standard Wi-Fi broadband connections at either 2.4 or 5 GHz are reaching various limitations in terms of its available bandwidth as the need for fast and efficient data transfer is increasing. To achieve larger throughput with increased bandwidth, with additional advantages such as enhanced security and reduced interference, Li-Fi technology is proving to be a contender for the last mile of data communication. Additionally, with its enhanced wireless capabilities Li-Fi provides the required connectivity developments to spearhead technologies such as the IoT and next-generation 5G networking. Li-Fi operates within the electromagnetic (EM) spectrum, and is therefore dependent on light wave technology. The following paragraph summarizes the EM characteristics of a Li-Fi connection.

1.7.1 Li-Fi as Visible Light Communication

Li-Fi shows potential to be integrated with fiber and radio wave communications for future gigabit networks. Its primary advantages all stem from the fact that Li-Fi uses light waves as opposed to traditional radio waves to transmit information. The underlying technology of Li-Fi is visible-light communication (VLC), broadcasting data by means of visible light by intensity modulation at frequencies between 400 and 800 THz of the available spectrum. Figure 1.4 depicts the EM spectrum and gives an indication of the wavelengths and frequencies of visible light which falls between the ultraviolet (UV) and IR spectra.

As shown in Fig. 1.4a, the EM spectrum is defined by its wavelength and frequency (from long-wavelength, low-frequency radio waves to short wavelength, high-frequency gamma rays). The available spectrum for Li-Fi, defined to oper-

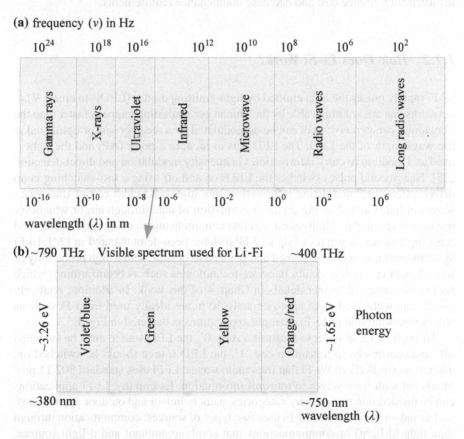

(a) frequency (v) in Hz

(b) ~790 THz   Visible spectrum used for Li-Fi   ~400 THz

Fig. 1.4 The EM spectrum (**a**) and the wavelengths and frequencies that can be utilized by Li-Fi for visible light communication (**b**). The approximate photon energies are also shown with respect to the wavelengths of the visible light

ate using visible light, is between approximately 400 THz (red light) and 800 THz (790 THz—violet light) as seen in Fig. 1.4b. These frequencies translate to wavelengths of between 380 and 750 nm, a reason why this technology is often referred to as nm-wave communication.

A distinguishing factor concerning VLC and Li-Fi is that VLC is a point-to-point data transfer technique, where Li-Fi describes a complete wireless networking system [6]. Therefore, conversely to the point-to-point infrastructure of VLC, Li-Fi provides bidirectional multiuser communication; enabling point-to-multipoint and multipoint-to-point data transfer [6]. Pulses of illuminated light are used to overcome bandwidth limitations currently hindering progress of network expansion using Wi-Fi operating in the spectrum below 10 GHz. In particular, Li-Fi can be considered an optimal solution to distribute a network connection—such as the internet—to densely populated areas, where the primary connection is split using a single VLC connection to end-users. In rural areas, this solution can eliminate constraints in infrastructure, reduce cost and decrease maintenance requirements.

1.7.2   How Does Li-Fi Work?

Li-Fi rapidly pulses the light emitted by light-emitting diodes (LEDs) to create VLC networks that are not noticeable by the human eye (pulsating happens faster than the human eye can perceive), but can be demodulated by a receiver sensor optimized at the wavelength of the LED. The LED acts as an access point (AP) and the light is used as a medium to carry information via intensity modulation and direct detection [25]. Nanosecond pulses switches the LEDs on and off using a fast-switching lamp driver, effectively modulating information on a high-bandwidth carrier (the wavelength of light) and allowing digital transmission of data through air, or whichever medium is applicable. High-speed wireless communications using single LEDs and reaching transmission rates of up to 3 Gbps have been demonstrated in [24]. Li-Fi systems must however be designed and planned to operate with a high probability of line-of-sight (LoS) and ideally integrate technologies such as beamforming, which will be discussed in further details in Chap. 4 of this book. In addition, relatively small cells with a radius of approximately 50 m are ideally used in Li-Fi systems [6]. A representation of Li-Fi example applications is depicted in Fig. 1.5.

Through Li-Fi, in order to transmit a zero '0', the LED source must be switched off, and conversely, to transmit a one '1', the LED source should be switched on. Therefore, similarly to Wi-Fi that uses radio waves, Li-Fi uses standard 802.11 protocols but with light waves to transmit information. Essentially, Li-Fi applications can be divided into two primary categories, namely indoor and outdoor applications.

For indoor applications, Li-Fi uses two types of sources: communication through data-light (d-light) or communications that combine ambient and d-light sources. Various light sources (d-light sources) are supported in VLC systems, including traditional LEDs at various wavelengths, micro-LEDs (μLEDs) which is an array of

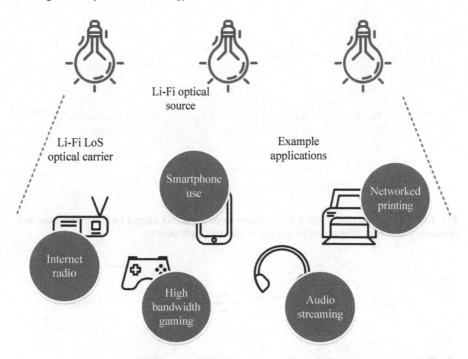

Fig. 1.5 A representation of Li-Fi example applications utilizing VLC to transfer information at high speed and low cost

typically gallium-nitride microscopic LEDs that form individual pixel elements as well as laser diodes [15].

Outdoor applications typically use sources such as streetlights, communication with satellites or transmission among moving vehicles. A typical functional diagram of a light source capable of transmitting data to a receiver, therefore a VLC system, is depicted in Fig. 1.6. There are several similarities between a VLC system and a traditional optoelectronic communication system as described in Lambrechts and Sinha [14].

As shown in Fig. 1.6, data modulated by a specified operating frequency is presented to a lamp driver, capable of switching on and off a light source at rapid intervals and supply adequate power to the source to generate light at an intensity required by the application. As the light pulses are generated in synchronicity with the modulated data, the visible light can travel through the medium, in most Li-Fi cases this would be air. Light detectors, or optical detectors, are placed typically in LoS of these transmissions, able to process incident light, and convert optical signals to electrical signals. If LoS cannot be achieved, signal degradation can occur, but essentially these systems are able to communicate if reflections from nearby objects are detected by the receiver. These electrical signals, typically a small current, is then amplified to a usable voltage by a transimpedance amplifier (TIA), and the

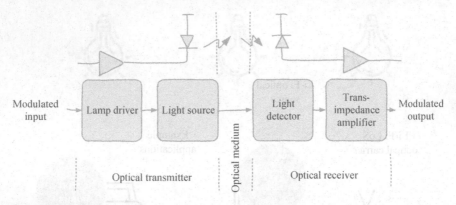

Fig. 1.6 A functional diagram of a VLC system modulating data using a visible light source and detecting transferred information by an optical sensor or light detector

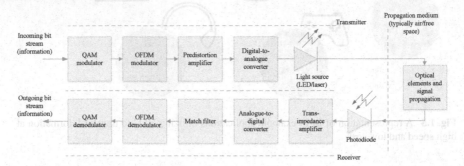

Fig. 1.7 A system-level functional diagram of a high-speed Li-Fi communications system, adapted from Jurczak [12]

data is extracted by additional supporting circuitry. Furthermore, Li-Fi uses certain commonly used modulation techniques, including

- single-carrier modulation which include on-off keying, pulse position modulation and pulse amplitude modulation,
- multi-carrier modulation which include variants of orthogonal frequency division multiplexing, and
- Li-Fi specific intensity modulation schemes such as color shift keying and color intensity modulation.

These modulation schemes are reviewed in further detail in Chap. 4 of this book. To expand on the simplified functional diagram of a Li-Fi communications system presented in Figs. 1.6 and 1.7 presents a system-level functional block diagram of a high-speed Li-Fi communication device, with respect to the components included in both the transmitter side and the receiver side of such a system, adapted from Jurczak [12].

As shown in Fig. 1.7, a high-speed Li-Fi system consists of various electronic and optical building blocks to realize an integrated system. From the transmitter, a bit stream of information is first modulated at the intensity level of the LED, using quadrature amplitude modulation (QAM) and encodes the incoming stream of data into symbols that will be loaded into the subcarriers. For high-speed optical wireless communication, multi-carrier modulation (MCM) is typically employed [12]. Therefore, parallel data streams are ideally modulated on an orthogonal frequency-division multiplexing (OFDM) scheme. In order to accurately reproduce the signal presented at the input, the linearity of the signal is improved by a pre-distortion amplifier before it is fed to the digital-to-analogue converter (DAC). At this point, the signal has been converted to an electric current and can be transmitted using optical elements towards the receiver. The weak received current is amplified to a voltage by the TIA and converted back to its digital form by the analogue-to-digital converter (ADC). A matched filter again linearizes the incoming signal to maximize the signal-to-noise ratio before it is demodulated through the OFDM and QAM modulators to retrieve the original information. Essentially, Li-Fi systems work on a similar foundation as most communication systems, with some exceptions, and is discussed in further detail in Chap. 4 of this book.

Li-Fi is considered a complete wireless networking system that offers bidirectional multi-user communication [12]. It achieves this with a wireless network of small optical cells and presents high spatial connection density with smooth handover. Additionally, Li-Fi presents several advantages and disadvantages in its use as a suitable replacement or as a complimentary technology in distributed broadband networking to devices, typically within a relatively short range from the transmitting light. These advantages and disadvantages are discussed in the following paragraphs.

1.7.3 The Advantages of Li-Fi

Advantages of this technology include:

- These networks are power-efficient and cost-effective communications as many homes and offices, even in rural areas, already have sufficient infrastructure to accommodate LED lighting. This same source of light can be converted to a VLC system to transmit data across short distances with enhanced wireless capacity to provide necessary connections to the IoT [6]. The same light source can be used to provide adequate light for a household or office whilst transmitting data since the frequency at which the light is pulsed on and off is faster than the human eye can observe.
- Li-Fi offers high availability in areas where advanced technologies are costly or unattainable, again since any source of light can be transformed to a d-light.
- It presents increased security of data transmissions since light cannot pass through opaque surfaces making the transmission impossible to intersect from adjacent areas separated by a physical wall or surface, making it easier to information

private. The relatively short-range of Li-Fi and its requirement of LoS for optimal communication can therefore be interpreted as an advantage, albeit these are
viewed as disadvantages in typical communication systems. VLCs are designed
to operate in specific environments that do not require long-range and non-LoS
(nLoS) communications.

- Li-Fi has an inherent advantage over traditional Wi-Fi with respect to the bandwidth it presents for data transmission to multiple users. The amount of bandwidth
  available in a typical Li-Fi system is enough to distribute data to a large amount of
  users and devices without suffering decrease in transmission speed. Typical bandwidth availability and characteristics such as power consumption are reviewed in
  Chap. 4 of this book.
- As a result of LEDs being the primary source of light for VLCs, the typical power
  consumption of these networks are low, especially considering that the light provided by source is also used as a traditional light source in dark areas or at night.
  In terms of actual wattage, in-depth comparison between traditional Wi-Fi routers
  and VLCs are presented in Chap. 4 of this book.
- Light signals are inherently immune to electromagnetic interference (EMI) and
  do not suffer losses or degraded performance from nearby sources emanating at
  comparable frequencies, as is the case with radio waves (and wired transmissions).
  Low susceptibility to EMI therefore translates to a capability of using multiple
  sources that transmit data and sensors receiving this data used in an area, effectively
  allowing dense network infrastructures to be implemented with little interference
  among them.
- Light-based data communication relies on the optical spectrum and therefore
  avoids having to design transmission protocols to operate in the already crowded
  radio-wave spectrum below 10 GHz [6]. The extremely large bandwidth availability within the light spectrum also future-proofs spectrum allocations as the
  probability of an over-crowded spectrum becomes less likely.

Although there are numerous advantages to Li-Fi and subsequently VLC-capable
networks, the technology (which is still in its infancy) also presents several limitations
and disadvantages that could impede its uptake as a standard technology.

1.7.4   The Disadvantages of Li-Fi

Some of the limitations and disadvantages of Li-Fi include:

- A general increase in overall power consumption in households and offices could
  be observed since the light sources are required to be on (albeit pulsing) in order
  to transmit data. This would effectively mean that the light source will be on even
  during the daytime when additional light—apart from the natural ambient light—is
  not required. Even if the power consumption of the light is minimal there will still
  be a significant amount of wasted energy from emitted light in already well-lit
  areas.

- Applications that implement Li-Fi are limited also by the necessity to have a light source on during transmissions. As an example, users will not be able to use network capabilities such as social media in personal spaces without having a light source illuminating the entire area, as could be the case when browsing the internet at nighttime when others prefer a dark room to sleep.
- As stated in the advantages of a communication that cannot penetrate walls or other opaque surfaces, this can also be seen as a disadvantage as the signal range between the light source and the receiver is limited. Receiving devices must be within LoS of the transmitting source (for optimal detection, although not necessary if reflections are detected) and deters users from moving around while accessing such a network. Through using high frequencies of several hundreds of THz for communication, the inherent relationship between frequency and distance defined by the Friis free space equation further limits the distances that data can be transmitted across.
- Although these networks are immune to EMI, other sources of light can still interfere with the primary signal. Ambient sources such as the Sun can interrupt and degrade the signals leading to inferior networking.
- An initial investment in such an infrastructure can be high, compared to traditional Wi-Fi technologies that have become cost-effective to install with relatively low complexity. Li-Fi systems would require a complete redesign in network infrastructure and would require significant planning and development in underdeveloped areas which could increase the cost above the feasible limits.

As indicated by the advantages and disadvantages of Li-Fi, it is possible to deduct that there would be applications that can benefit from this technology, although it will possibly not manifest as a replacement to traditional radio waves or wired connectivity. Hybrid systems [5] can combine VLC and radio frequency (RF) communication systems to provide pervasive coverage through heterogeneous solutions that take advantage of the most promising characteristics of both systems and design the infrastructure accordingly.

1.7.5 Applications of Li-Fi

With respect to the last mile, this technology offers an intriguing set of characteristics that make it a strong contender for delivering connectivity to areas and infrastructures that may otherwise not be suited for traditional distribution. Some examples of application of Li-Fi where the advantages of this technology outweigh the limitations are given below.

- Wireless communications and illuminations through implementing Li-Fi in street-lights. This application can be especially useful in dense urban environments, rural areas, developing countries and under-developed infrastructures where large amounts of people can access networking capabilities and the internet in hotspots with traditionally restricted and limited access. High bandwidth capabilities can ensure that large amounts of users can access a single network with little degradation in performance and user experience.

- In EMI-sensitive environments, for example in aviation and in the healthcare sectors, Li-Fi provides an alternative to technologies that can interfere with critical devices. In the healthcare sector, equipment such as magnetic resonance imaging is susceptible to radiation from interfering signals and also emanate interfering signals to other equipment. Commercial and private airplanes in the aviation sector could implement Li-Fi as an alternative to transmit in-flight entertainment and internet connectivity to passengers. This will not only reduce potential EMI, but also reduce the complexity of current wired systems and inevitably reduce the weight overheads introduced by outdated systems. In both the aviation and healthcare sectors, sufficient and efficient lighting is already a priority and would have a minimal impact on the illumination standards already implemented.

- Li-Fi can be incorporated into headlights of vehicles, where modern cars already have always-on LED running lights, to communicate among vehicles and act as an early warning system when vehicles enter into a dangerous proximity with each other. Autonomous vehicles can also communicate through headlights and alternative lighting, effectively complementing (or replacing) more expensive radio detection and ranging (radar) technology.

- Areas that need to be permanently illuminated, typically hazardous environments such as petrochemical plants, would benefit from introducing Li-Fi in its already-implemented illumination infrastructure. Cost-effectiveness through eliminating the need to include radio wave transmissions is advantageous in such scenarios, with an additional benefit of lowering the risk of EMI among sensing devices that are typically extremely sensitive when detecting chemical or radiative changes in the immediate environment.

- Li-Fi can provide media-rich information to users in scenarios where an object (for example an art piece in a museum or an item in a retail store) is individually illuminated to showcase the object. Interaction with these objects can therefore occurs through a Li-Fi network and the light source only needs to be on during times when the object is exhibited.

- Finally, Li-Fi has a unique capability of permitting underwater communication, a feat that radio waves are unable to contend with. Light does get absorbed by moisture, but to a lesser extend when compared to radio waves. Also, wired communications such as long-haul networks under the ocean are expensive and has a limited lifetime due to corrosion, therefore strategically-placed Li-Fi APs can potentially complement traditional methods of distributing information underwater.

Technology such as Li-Fi can also assist communities and rural areas to become connected, thriving and sustainable if implemented correctly and maintained throughout its lifetime. Apart from rapid urbanization in developing nations, technology limitations in developing worlds have different restrictions that impede its uptake. The following paragraph describes how Li-Fi can be implemented in rural communities, where other traditional technologies have been deemed not viable or have additional limitations that encumber them.

1.7.6 Li-Fi Networks in Rural Communities and Its Expected Market Growth

Connectivity in rural communities must be addressed by infrastructures that warrant improvements for large portions of the populations that would otherwise not have access. Furthermore, technologies such as Li-Fi can be combined with sustainable energy-harvesting devices for example solar panels to contribute the necessary means for uninterrupted power supply to these system. In rural communities such as in India and many African countries, availability of power, not even considering stable and uninterrupted power, is a privilege and could be considered a caveat when introducing new technologies. Renewable energy such as photovoltaic cells in solar panels as well as wind and hydro-energy allow sustainable means of providing these sources of energy. Li-Fi enabled solar panels have already been developed as a solution for rural communities in India to ensure connectivity irrespective of the location of the proposed system. There are several advantages to these Li-Fi enabled solar panels, including:

- Cost effective investments—by merging two up and coming technologies (solar power and high-speed connectivity through Li-Fi), subsidies from government and a drive to lower installation cost can drastically reduce the cost of such systems; providing energy and connectivity in a single solution.
- No requirements for regulated power supplies from local utility distributors through solar panel networks.
- Li-Fi networks can be built on the backbone of existing optical fiber and copper-based networks where available, typically in urbanized areas.
- Through solar panels, energy can be fed back to the grid, making these investments much more feasible and sustainable for investors, since funds can be further subsidized in this way.

The global Li-Fi market is also set to grow significantly within the next few years. Market forecasts of a compound annual growth rate of over 55% from its US$ 2.4 billion market share at the end of 2018 are estimated by Business Wire, a Berskhire Hathaway Company. Various research and development centers in North America are already spearheading the market and contributing a significant share of the revenue of this market. Large electronic markets in Asia, specifically in countries such as China and Japan, are also expected to benefit from Li-Fi as it becomes more adopted worldwide. In some emerging regions, local governments are promoting the usage of LED lights for cost-effective energy usage, beneficial for both the consumer and the suppliers. Essentially, technologies such as Li-Fi are expected to witness a higher growth compared to more niche technologies, primarily due to the advantages it brings for both investors and users. In the Middle East and Africa, Li-Fi is also expected to seize a high portion of the connectivity market, currently dominated by Wi-Fi and cellular. Although cost effective Wi-Fi solutions can be put in place in these countries, cellular in particular are still very expensive in these regions and do not offer cost-effective and sustainable means of connectivity to low-income families. The internet has been accepted globally as the most important enabler of social development and economic growth, but with large portions of our population that do not have access to connectivity, Li-Fi aims to address these issues.

1.7.7   Li-Fi as the Last Mile Access

A fundamental step in driving light wave connectivity is to use Li-Fi as a widespread alternative in last mile communication. Light poles should essentially become a key element of smart city connectivity as well as infrastructure [23]. These light poles should in turn be connected to a larger bi-directional communications network through technologies such as fiber optics or PLC. Albeit a significant potential contributor in creating the smart cities of the future, Li-Fi must also be seen as a strong contender in providing access to rural communities and geographically difficult to reach areas.

Another high-bandwidth communication strategy that is gaining traction and considered a viable alternative to the last mile is millimeter-wave (mm-wave) backhaul technology. Especially in rural areas, it can be complex and sometimes not economically viable to provide traditional copper- or fiber-based backhaul access to individual cells that distribute network access to the end-user. An increase in the maturity of mm-wave communication is enabling opportunities to provide high-speed wireless backhaul connection to cells (the term *cells* is discussed in the following section). In leiu of these advancements in mm-wave technology, using it as a backhaul has several advantages, presented and discussed in the following section.

1.8   Millimeter-Wave Backhaul Technology

Mobile connectivity has grown significantly in recent decades and the broadband requirements of these networks are increasing as more users are added and higher-bandwidth content is consumed. As these bandwidth requirements become more intensive, broadband service providers are being forced to update traditional copper-based (twisted pair and coaxial cable) to infrastructures that are capable of handling high network traffic demands, such as optical fibers. The issue with this approach, however, is that upgrading existing infrastructures to fiber or adding new ones is a time-consuming and expensive exercise. Limitations in rural areas also prohibited these upgrades in many circumstances. In terms of the last mile, fiber is in many ways an ideal solution, but again looking at developing and rural areas, not always attainable. A high-speed wireless alternative is in many cases an ideal choice and made possible through mm-wave communication.

1.8.1 What Is mm-Wave?

In the EM spectrum, mm-wave refers to frequencies between 30 and 300 GHz. The wavelength λ in m of a frequency f in Hz is determined by the equation

$$\lambda = \frac{c}{f} \tag{1.1}$$

where c is the speed of light at approximately 299,792,458 m/s. Through [1], the frequency with value 300 GHz relates to a wavelength of 1 mm and 30 GHz relates to 10 mm (or 1 cm), hence the term mm-wave frequencies. In the EM spectrum, mm-wave frequencies fall between IR and microwave signals, as presented in Fig. 1.8.

These mm-wave frequencies, between IR and microwave frequencies as shown in Fig. 1.8, are still considered radio waves and have the potential to provide various solutions for the growing bandwidth demands in modern communications. Semiconductor materials such as indium phosphide, gallium nitride and gallium arsenide are making it possible to manufacture active components (such as transistors) in

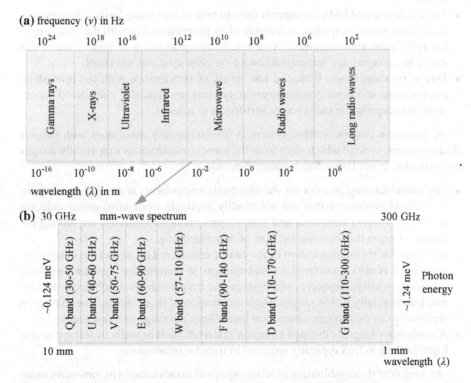

Fig. 1.8 The EM spectrum (**a**) and the wavelengths and frequencies that defines mm-wave (**b**). The approximate photon energies are also shown with respect to the wavelengths of mm-wave

complementary metal-oxide semiconductor (CMOS) processes that are capable of switching at high speeds required in a mm-wave communication system.

There are three primary considerations of a mm-wave radio signal that make it attractive to use in a multitude of applications, typically short-range applications. These considerations include its

- bandwidth capabilities, typically in the unlicensed spectrum,
- short wavelength propagation and therefore its footprint of electronic devices such as antennas, and
- interaction phenomena with the immediate atmosphere.

Taking into account these considerations, mm-wave communication presents several advantages and disadvantages when used as a primary means of transferring information between a transmitter and receiver. Advantages of mm-wave communications include:

- Data transfer speeds in excess of 1 Gbps can be achieved using mm-wave radio waves, a requirement in modern data transfer applications such as high definition video streaming, or serving numerous users high-speed and low-latency network access.
- The electronic and EM components used to realize mm-wave-based circuitry typically have small footprints, as a result of its proportionality with the wavelength (as is the case with antennas). As a result, miniaturization of these circuits can easily be achieved and integrated system-on-chip solutions are small.
- Due to its short range, there are low levels of interference with the immediate environment and these communication systems are typically considered secure since intercepting the radio waves is difficult to achieve.

A mm-wave communication system is also inherently associated with various disadvantages, most of which stem from the same considerations that the advantages are extracted from. These disadvantages include:

- The manufacturing process for the electronic components are costly and require specialized equipment that are not readily available from most semiconductor foundries. Process variations and scarce semiconductor materials are among the primary factors that increase the cost of manufacturing.
- Although short communication range can be considered an advantage in terms of security, it is also considered a disadvantages in commercial applications since it requires multiple repeaters and high-power transmitters to reach economically and commercially viable ranges. High attenuation of these signals due to oxygen (referred to as the oxygen-absorption band) limit the range to a few meters.
- Attenuation of signals through common materials such as walls or foliage is also high, leading to LoS typically required to transfer information.

As a result of the combination of advantages and disadvantages of mm-wave radio waves, its current lack of maturity still make these implementations less widespread; however, the advantages that can be gained through its implementation are driving manufacturers to pursue its maturity and provide cost-effective solutions for various high-speed applications.

In the argument for mm-wave backhaul technology, 5G is spearheading the drive towards next-generation and cost-effective solutions for last mile access. Backhaul technology in wireless (cellular) technology refers to the section of the network that connects the cellular radio station to the core network. These links are typically high-bandwidth and low latency connections and include fiber, copper-based, microwave or in certain circumstances, satellite links [11].

1.8.2   5G Technology Integration with Cells

Next-generation wireless technologies such as 5G builds on previous generations with improvements in bandwidth and transmission time intervals. Through a combination of innovative new radio interface concepts and advanced multiple-input-multiple-output technology, 5G achieves wider bandwidth at higher frequencies. Performed 5G tests utilize frequencies around 15 GHz, compared to 3G and 4G operating within the 2.6 GHz band [19]. This higher frequency results in shorter range which drives its potential applications in densely populated areas where multiple base stations can be deployed to deliver last mile access.

5G is capable of serving multiple users high-bandwidth content and can be deployed through scalable network infrastructures utilizing large numbers of small cells (for example a 3G or 4G small cell) and fixed broadband APs (such as a Wi-Fi AP). One implementation is, to deploy mm-wave broadband solutions that provide high-bandwidth data transmission links to small cells and fixed broadband APs. A fundamental concept of cellular networks is to divide the space covered by the service in limited areas, typically achieved by cells [9]. Each of these cells is individually served by a base station that consists of a transmitting and a receiving antenna, wired to a central network and capable of managing other base stations as well as the flow of data and information across the network. 5G is capable of providing high-bandwidth information to users, provided its distribution is of high quality with low loss and little signal degradation along its path to multiple users. In this sense, 5G provides an optimal backbone of high-speed data where the last mile is tasked with distributing each user with his or her share of the available bandwidth; based on the capacity needed by each user. The premise of small cells, used to distribute large bandwidth information transfer at relatively short range, is discussed in the following paragraph.

1.8.3   Small Cells

A small cell is a low-powered radio access node with a relatively short range or a few meters up to several kilometers. It is therefore essentially a miniature base station of a cellular site to increase the edge data capacity, speed and overall network efficiency of the cellular network. The primary cellular signal is received from the macrocell and distributed by small cells. Small cells are categorized in subsets based on their range and power delivery into (from smallest to largest) femtocells, picocells and microcells—all compared to the mobile macrocell. These cells can be densely distributed in urban areas and are targeted at indoor and outdoor locations, depending on the application, cell size and the overall infrastructure. Ideally, small cells can be implemented in rural areas that are far away from the primary macrocell, and distribute network connectivity to multiple users within a community. Table 1.4 summarizes the characteristics such as where each type of small cell is typically implemented, the number of users it typically serves and other characteristics relating to its power and bandwidth capabilities.

As shown in Table 1.4, the typical premise is that for larger cells (therefore from the smallest femtocell to the larger macrocell), the number of users, output power, cell radius and bandwidth increases. Therefore, smaller cells like femtocells are typically used indoors to serve a group of people within a community with high-quality network connections. The transmit range of these cells are limited and the application/environment of its implementation is specifically chosen and designed for. As high-bandwidth 5G cellular is evolving, small cells play a critical role in this transition. The small cells provide advantages in infrastructure development such as

- an increase in data throughput to individual users,
- a reduction in the cost for service by reducing the necessity for additional macrocells,
- a reduction in processing and power requirements of end-user devices if signal strength is stronger, and
- to allow remote rural areas to gain access to the core network at lower investment costs.

Furthermore, mm-wave 5G cellular signals will require multiple small cells to distribute the signals since these signals are typically not capable of penetrating

Table 1.4 Summary of the types of small cells and the relevant characteristics associated with each

Characteristic	Femtocell	Picocell	Micro-/metrocell	Macrocell
Indoor/outdoor	Indoor	Indoor/outdoor	Outdoor	Outdoor
Number of users	4–16	32–100	200	>200
Max output power	20–100 mW	250 mW	2–10 W	40–100 W
Max cell radius	10–50 m	200 m	2 km	10–40 km
Bandwidth	10 MHz	20 MHz	20/40 MHz	60–75 MHz
Technology	3G/4G/Wi-Fi	3G/4G/Wi-Fi	3G/4G/Wi-Fi	3G/4G
MIMO	2×2	2×2	4×4	4×4
Backhaul	DSL/cable/fiber	Micro-/mm-wave	Fiber/microwave	Fiber/microwave

walls and have limited range, a function of its short wavelength. In Chap. 5 of this book, small cells and the capabilities of each variant is reviewed in detail, to determine the optimal infrastructure for a completely heterogeneous network. A heterogeneous network is a combination of small cells that operate in conjunction to provide coverage to as many users as possible and maintain network stability, availability and speed. Millimeter-wave backhaul technology with 5G can thrive in indoor environments where there is an increasingly higher ratio of total mobile data consumed, with wireless access to the internet being offered to larger groups of users.

An implementation that is commonly used to serve multiple users in an indoor environment is a distributed antenna system (DAS). Capacity scaling of distributed sectors can be achieved by deploying multiple base stations and reducing the number of users for each sector. The following paragraph describes the essential configuration considerations of a DAS.

1.8.4 Distributed Antenna System

A DAS is an integrated network of antennas that can send as well as receive mobile cellular signals on the licensed frequencies of a specific carrier. In its simplest form, a DAS consists of two vital components. These components are the

- signal source that is responsible for generating the cellular signal, for example the 5G signal and
- a distribution system that is responsible for distributing the signals throughout the (typically indoor) environment.

For a DAS there are several types of signal sources and distribution systems. Typical signal sources include off-air sources generated by antennas (also referred to as repeaters) typically on the roof of the indoor environment, an on-site base transceiver station (BTS) or small cells, as discussed in the previous paragraph. Off-air signal sources comprise of a donor antenna that receive a weaker signal from a remote base station and amplify the signal towards the distribution system. The off-air implementation is typically the most cost-effective but is heavily dependent on the quality of the received signal. The donor antenna is only capable of amplifying the quality it sensed and cannot improve the quality (therefore the bandwidth) of the signal. In essence, this source only acts as a device that extends the incoming high-bandwidth signal.

An on-site BTS is a technology used in cellular towers to generate the signal itself. A common implementation of a BTS requires a high-bandwidth connection between the BTS of the cellular carrier and the core network signal that is to be distributed by the DAS. For high-volume data consumption, multiple instances of the BTS could be connected to various cellular carriers to alleviate the effective load of each BTS.

Finally, the use of small cells has become more prominent in modern implementations due to its modularity and relatively low complexity of installation. The small cells create a direct and secure tunnel to the network established and managed by the cellular carrier using a typical internet connection of preferably high-bandwidth capabilities. Each small cell requires a dedicated connection to the internet which is considered a limitation of this technology; however, it modularity and relatively low cost outweigh this limitation and therefore its use is becoming more prevalent to provide multiple users with network connections. Additionally, all these alternative implementations can be combined and its overall interoperability further extend the capabilities and application-specific benefits of DAS. Table 1.5 summarizes the advantages and disadvantages of off-air, BTS and small cell configurations used as signal sources in a DAS.

As shown in Table 1.5, the advantages and disadvantages of each DAS implementation is heavily dependent on the environment in which the solution is to be incorporated. A general trade-off in cost, capacity and complexity must be considered to ensure that the proposed solution is viable in a specific scenario. For this reason, planning typically consists of combining off-air, BTS and small cells making the overall implementation of the DAS (to essentially optimize the last mile) unique for different applications. The primary goal of the DAS is to amplify an appropriate signal, distribute it and rebroadcast the high-quality network signal throughout an indoor environment (or in certain outdoor scenarios). Furthermore, a DAS consists of four primary categories of implementations. These categories are

Table 1.5 Advantages and disadvantages of off-air, BTS and small cell configurations used as signal sources in a DAS

DAS signal source	Advantages	Disadvantages
Off-air (donor antenna)	Rapid deployment and little carrier involvement	Performance degradation if source signal is low
	A low-cost solution	Not able to increase obtained capacity
	Can be used with numerous carriers	Signal optimization for different carriers can be complex
Base transceiver stations	Superior performance compared to other solutions	An expensive solution compared to off-air and small cells
	Ability to scale capacity based on user requirements	Deployment times can be excessive
		High power requirements that require additional cooling (costly)
Small cells	Lower cost with high levels of modularity	Scaling implementations to increase capacity can be complex
	Rapid deployment possible	Relies on a high-quality primary internet connection
	Can be used by large numbers of users if properly scaled	Can be limited if carriers do not allow for small cell integration

- active,
- passive,
- hybrid or
- purely digital.

An active DAS is capable of converting an analogue RF transmission that it receives from the signal source to a digital representation of the signal. It therefore requires a high-bandwidth ADC that can perform the conversion on either a signal from a single carrier, or from multiple carriers at once, effective increasing the cost, complexity and bandwidth requirements of the ADC. Figure 1.9 is a simplified representation of an active DAS.

In Fig. 1.9, the digitized signal is then distributed by the DAS over a high-speed and low-loss medium such as fiber optics or an Ethernet connection. To its advantage, an active DAS is easily expandable and the last mile cables can be of long length with limited signal degradation based on the infrastructure (with low-cost repeaters used to amplify the signal along its path). The distributed signals can also easily be connected to a wireless implementation such as Wi-Fi to further distribute the wireless signal to multiple users. An active system is however an expensive solution as it requires amplification and conversion circuitry throughout the network and the remote units require its own dedicated power, further increasing the overall cost.

Alternatively, a passive DAS does not require conversion of the RF signal and distribution is performed by low-cost coaxial cables, various splitters, taps and couplers. Figure 1.10 is a simplified representation of a passive DAS.

In the passive solution shown in Fig. 1.10, there is however a significant loss in signal power and therefore integrity as a function of distance between the receiver and the signal source. The maintenance of these implementations is relatively straightforward and to its advantage, no additional equipment is required to expand the solution to multiple carriers. To its disadvantage, such a solution requires a precise link budget to ensure that the capacity and signal integrity is high enough to service all

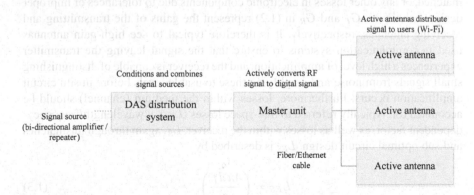

Fig. 1.9 A simplified representation of an active DAS where the master unit converts an incoming analogue RF signal to a digital representation and distributes the signal through a fiber optic link or Ethernet

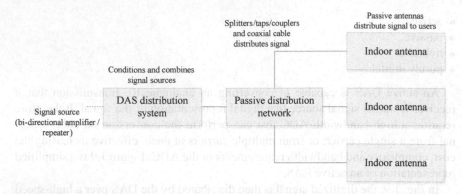

Fig. 1.10 A simplified representation of a passive DAS where the signal from the source is distributed through a passive medium such as a coaxial cable and passive components such as splitters, taps and couplers

users. The link budget refers to all of the gains (through amplification) and losses (as the signal propagates through various mediums) between a transmitter and receiver in a typical communications system. A common equation used to determine the overall link budget in a communications system, is

$$P_R = P_T - L_T + G_T + G_R - L_{FS} - L_R \qquad (1.2)$$

where P_R is the power received by the receiver, for example the strength of the network signal received by the device carried by the user. This received power is a function of various parameters and characteristics within the communication system, including the power of the transmitted signal, P_T, and the losses associated with the transmitter. These losses are internal to the design and the components used by the transmitter and can be experienced where connectors are present that are not properly matched, or any other losses in electronic components due to tolerances or improper design techniques. G_T and G_R in (1.2) represent the gains of the transmitting and receiving antennas, respectively. It is therefore typical to see high-gain antennas used in communication systems to ensure that the signal leaving the transmitter experiences a high level of amplification, and the receiver is capable of distinguishing small signals from noise and amplifying these to usable levels before in situ circuit amplification occurs. Furthermore, losses within the medium (channel) should be accounted for, typically referred to free-space losses (L_{FS}), a wavelength/frequency-dependent factor, as well as losses within the receiver, L_R, again through mismatches and sub-optimal circuit design. L_{FS} is described by

$$L_{FS} = \left(\frac{4\pi d f}{c}\right)^2 \qquad (1.3)$$

where d is the distance between the transmitter and the receiver, in meters, and f is the frequency of the signal. The wavelength of the signal, λ in meters, can be deduced from simply determining the value of c/f. The link budget and an analysis on typical free-space losses in communications systems, and its relevance to the last mile, is described in further detail in Chap. 3 of this book.

To mitigate some of the disadvantages of both active and passive DAS, it is common practice to realize hybrid solutions that combine the favored characteristics of active and passive systems. Figure 1.11 shows a hybrid solution that uses both active and passive components in signal distribution.

In a hybrid DAS as shown in Fig. 1.11, remote radio units (RRUs) use either fiber optic or Ethernet connections to connect to the main unit and distributes the signal through the indoor environment. The main unit convert the RF signal from its analogue form to a digitized representation and the RRUs are fed to multiple antennas that operate independently. Since less RRUs are required in a hybrid solution, the overall cost is typically less compared to a purely active DAS. Additionally, there are no limits in the length of the cables used in the digital backbone of the solution. These systems are still more expensive that a purely passive solution, also requires link budgets for each front-end and installation can become complex since both fiber/Ethernet as well as coaxial cabling is required.

Another solution, albeit used less often that active, passive or hybrid, is a purely digital DAS. In a digital DAS, the baseband unit which acts as a type of BTS, can communicate directly with the master unit through remote units and essentially no conversion from analogue RF signals to digital signals is required. In theory, this type of DAS would offer the lowest cost solution and overall lowest complexity, but its caveat is the large number of competing standards among carriers that make its deployment practically impossible. Furthermore, employing mm-wave backhaul networks require new hardware at the base station and hence global standardization before full commercialization can be achieved [20].

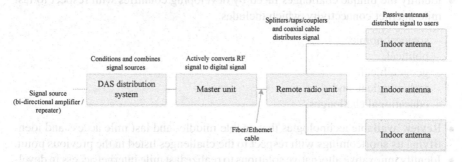

Fig. 1.11 A simplified representation of a hybrid DAS where remote radio units connect the main unit and distributes the signal through an indoor environment

1.9 Conclusion

Socioeconomic prosperity and thriving businesses are becoming more dependent on internet connectivity, but the digital divide still sees nearly 4 billion people not having access to the internet. There are numerous challenges that must be overcome to provide access to more people across the globe, with the ultimate goal to provide internet access from backhaul and base stations to end-user devices. This can be achieved through a combination of complimentary technological advances and partnerships in innovative business models, governmental support and training generations of skilled workers.

To address the issues plaguing last mile access, there is not a single solution and requires intervention from multiple sectors. Community-based innovation centers, travelling roadshows, training for low-income individuals, combining formal engineering and innovation training and improving computer literacy skills globally are among the potential endeavors to address challenges, specifically in developing countries.

This book aims to provide a combination of topics to the reader to highlight not only the concerns in developing countries, but also provide potential solutions based on a strong technology motivation, current successes achieved in delivering last mile internet access and proposed policies to sustain development and encourage innovation. The Global South has an ideal opportunity of being not only a part of the fourth industrial revolution, but also in growing into a recognized leader. Overcoming the digital divide does not only have to be seen as a retrospective correction but should lead the way for other more developed countries in implementing lower-cost, efficient and sustainable last mile solutions. Developing countries have unique challenges they are facing and can often be implemented even in circumstances where the same challenges do not exist.

Therefore, the aim of this book is to:

- Identify the unique challenges faced by developing countries with respect to last mile internet connectivity, which includes

  - socioeconomic,
  - political,
  - geographical,
  - technological, and
  - educational challenges.

- Review available technologies that enable middle- and last mile access and identifying its shortcomings with respect to the challenges listed in the previous point.
- Identify innovative alternative solutions to realize last mile internet access in developing, rural and/or geographically difficult to reach environments.
- In-depth analysis of the proposed solutions with focus on the technologies driving these alternatives; providing the reader with the tools and knowledge to identify such solutions and the theoretical background to understand its advantages and/or limitations.

- Provide examples and references of successful implementations of novel and traditional solutions with case-specific specifications and challenges.

Furthermore, this book also reviews key technologies that offer sustainability to last mile solutions in terms of deliver energy that is renewable, clean and can be used in areas where power from the main grid is either unreliable or not available, another challenge that is prominent in developing countries and overlook in many developed countries. In its entirety, this book can provide the necessary background, theoretical understanding and longevity of last mile internet access in challenging environments.

References

1. Aijaz R (2017) Smart cities movement in BRICS. Global Policy and Observer Research Foundation, London
2. Bull G, Garofalo J (Sept 2004) Internet access: the last mile. Learn Lead Technol 32(1):16–21
3. Faltings B, Jingshi J, Jurca R (2014) Incentive mechanisms for community sensing. IEEE Trans Comput 63(1):115–128
4. Fernández RA, Zubelzu S, Martinez R (eds) (2017) Carbon footprint and the industrial life cycle: from urban planning to recycling. Springer International Publishing AG. ISBN: 978-3-319-54984-2
5. Gupta A, Garg P, Sharma N (2017) Hybrid LiFi—WiFi indoor broadcasting system. In: IEEE 28th annual international symposium on personal, indoor, and mobile radio communications (PIMRC), Montreal, Canada, 8–13 Oct 2017
6. Haas H, Yin L, Wang Y, Chen C (Mar 2016) What is LiFi? J Lightwave Technol 34(6):1533–1544
7. Holmner M, Britz JJ (June 2013) When the last mile becomes the longest mile: a critical reflection on Africa's ability to transform itself to become part of the global knowledge society. Innovation (46):117–134
8. Hussian R, Sharma S, Sharma V, Sharma S (2013) WSN applications: automated intelligent traffic control system using sensors. Int J Softw Comput Eng 3(3):77–81
9. Iannacci J (2017) The future 5th generation (5G) of mobile networks: challenges and opportunities of an impelling scenario. In: RF-MEMS technology for high-performance passives, Chap. 3
10. Irandoust K (2016) The 5 biggest emerging markets in the IoT. Retrieved 5 Jan 2019 from http://itnext.io
11. Jaber M, Imran MA, Tafazolli R, Tukmanov A (Apr 2016) 5G backhaul challenges and emerging directions: a survey. IEEE Access (4):1743–1766
12. Jurczak C (2017). LiFi: enlightening communications. Lucibel White paper. Retrieved 5 Apr 2018 from http://www.lucibel.io
13. Kreische F, Ullrich A, Ziemann K (2015) Internet of things. Using sensors for good: How the internet of things can improve lives. Published by Deutsche Gesellschaft für Internationale Zusammenarbeit (GIZ) GmbH, Bonn
14. Lambrechts JW, Sinha S (2016) Microsensing networks for sustainable cities. Springer International Publishing. ISBN: 978-3-319-28357-9
15. Li X, Hussain B, Wang L, Jiang J, Yue CP (Jun 2018) Design of a 2.2-mW 24-Mb/s CMOS VLC receiver SoC with ambient light rejection and post-equalization for Li-Fi applications. J Lightwave Technol 36(12):2366–2375
16. Mennecke BE, West LA (Oct 2001) Geographic information systems in developing countries: issues in data collection, implementation and management. Iowa State University, Digital Repository

17. Miazi NS, Erasmun Z, Razzaque A, Zennaro M, Bagula A (2016) Enabling the internet of things in developing countries: Opportunities and challenges. In: Proceedings of 2016 international conference on informatics, electronics and vision (ICIEV), pp 564–569. https://doi.org/10.1109/iciev.2016.7760066

18. MTC (2016) Metropolitan Trading Company (SOC) Ltd: 2015/16 integrated report. Retrieved 27 Apr 2018 from https://mtc.ca.gov

19. Neira E (2014) The latest update on 5G from IEEE communications society. Retrieved 28 Apr 2018 from http://www.comsoc.org

20. Pi Z, Choi J, Heath R (Apr 2016) Millimeter-wave Gbps broadband evolution towards 5G: fixed access and backhaul. IEEE Commun Mag 54(4):138–144

21. South African Cities Network and University of Witwatersrand (2017) BRICS cities: facts and analysis 2016. South African Cities Network, Johannesburg. ISBN: 978-0-620-72371-8

22. Stankovic JA, Wood AD, He T (2011) Realistic applications for wireless sensor networks. Theoretical aspects of distributed computing in sensor networks. Springer, Berlin, Heidelberg, pp 835–863

23. Sutton M (2018) LiFi set to light up the last mile. Retrieved 19 Apr 2018 from http://www.itp.net

24. Tsonev D, Chun H, Rajbhandari S, McKendry J, Videv S, Gu E, Haji M, Watson S, Kelly A, Faulkner G, Dawson M, Haas H, O'Brien D (Apr 2014) A 3-Gb/s single-LED OFDM-based wireless VLC link using a gallium nitride μLED. IEEE Photonics Tech Lett 26(7):637–640

25. Wu X, Safari M, Haas H (Dec 2017) Access point selection for hybrid Li-Fi and Wi-Fi networks. IEEE Trans Commun 65(12):5375–5385

26. Zennaro M, Pehrson B, Bagula A (Oct 2008) Wireless sensor networks: a great opportunity for researchers in developing countries. In: Proceedings of WCITD2008 conference, vol 67. Pretoria, South Africa

Chapter 2
Limitations of Last Mile Internet Access in Developing Countries

Abstract The digital divide negatively impacts socioeconomic growth, especially in emerging markets where numerous additional challenges and issues exist that also threaten inclusivity through widespread broadband internet access. Mitigating limited or no access to the internet and decreasing the disconnected population in emerging markets is not receiving the priority and urgency it deserves and its importance is often underestimated. Developed countries are not always aware of the additional issues that are dealt with in emerging markets on a daily basis, albeit these challenges might also exist to a lesser extent, in some form or another, in the developed world. Issues such as affordability of broadband internet measured against per capita gross-domestic-product, a lack of infrastructure to distribute the internet, regulatory challenges and incompetence and mismanagement of spectrum allocation plague internet affordability and distribution in emerging markets. Furthermore, more pressing issues such as corruption, famine, inequality, unreliable electricity, unemployment and education are among the serious issues that must be dealt with in emerging markets before internet access can be prioritized. This chapter researches the numerous challenges in emerging markets that are not technology-related but has a severe effect on the accessibility of the internet and the integration of last mile solutions in developing countries. Such a methodology benefits the research on bridging the digital divide in emerging markets.

2.1 Introduction

Technology advances are rapidly transforming how we work, communicate, access information, and educate ourselves. Current trends in information and communications technology (ICT) such as artificial intelligence, the internet-of-things (IoT), and cryptocurrencies are powerful tools to create social economic wellbeing without being geographically bound. However, connectivity is not a given for more than 4 billion people worldwide. Especially in developing countries, and in rural areas, there are numerous people that cannot access the internet for various reasons, economically, socially, or politically, and these communities will be left behind if connectivity is

© Springer Nature Switzerland AG 2019
W. Lambrechts and S. Sinha, *Last Mile Internet Access for Emerging Economies*,
Lecture Notes in Networks and Systems 77,
https://doi.org/10.1007/978-3-030-20957-5_2

not expanded. The fourth industrial revolution (Industry 4.0) is currently happening, and connectivity is crucial to take part in this movement.

Broadband internet has become an integral part for a new generation to

- start and grow small businesses,
- advance industries such as education, health care, agriculture,
- develop medium- and large businesses, and
- spearhead humanitarianism.

The *disconnected* population is often referred to as the digital divide, and to mitigate this, private sector investments and public sector support need to be combined with the common goal to deliver internet access to the citizens who are prevented from accessing these services. Extending the reach of the internet to include all, or at least, most of the developing world is an underappreciated means of reducing poverty. Vast scales of economic activity and employment opportunities have been created worldwide through the internet, and these statistics will only increase in the near future as the internet becomes even more integral; typically through electronic commerce (e-commerce), big data, and social media.

An identified problem with extending the internet to bridge the digital divide is a tendency of countries and states to implement an outdated macro policy perspective. Numerous decision makers are concerned about competition policies and equal access as an effect of local companies such as service providers or telecommunication operators. Another problem is the affordability of broadband internet and a global shift is required that sees selling access to the internet being replaced by other services, offered via the internet. This is off course not a realistic expectation, for now, since infrastructure costs to serve users with the internet is costly. To remove the affordability barrier, connectivity must be distributed at lower cost and more efficiently, through modernization of infrastructures and financial backing, ideally from governmental entities that realizes the importance of socioeconomic growth through connectivity. Economic diversification and structural transformation must be reshaped throughout the world as digital technologies are spearheading production processes and business models. Many developed and developing countries have recently experienced declines in the importance of manufacturing in their respective economies, a development that could lead to lower productivity, employment, and income opportunities [19].

The aim of this chapter is not to provide the statistics of internet connectivity in developing countries and lead the reader to realize the problems that emerging markets are facing in this regard. The aim of this chapter is to identify challenges that prevent internet connectivity from becoming ubiquitous in developing countries and rural areas. The theme of this book is to provide solutions to bring internet connectivity using emerging technologies to these groups of people, whereas this chapter identifies and highlights the challenges that impede it. Industry 4.0 is an integral part of modern society and will be responsible for multiple endeavors that head socioeconomic development. The chapter therefore starts with a brief overview of Industry 4.0 and its defining characteristics.

Following the introduction to Industry 4.0, this chapter identifies the issues in developing countries that prevent Industry 4.0 through stifling internet access for all. The first challenge that is presented is concerning the affordability of the internet in developing countries. Licensing fees and high prices for broadband data, and the reason for it, are identified. Subsequently, the lack of infrastructure in these areas are discussed, identifying why there is a gap in infrastructure and ways to diminish this. Another major challenge in developing countries is monopolies and anticompetitive conduct, commonly from governmental level, which makes access to the internet elusive and expensive. Macro policies and regulatory challenges follows this section, as they are closely related. Corruption in developing countries, and in developed countries for that matter, also stifle the growth of the internet in these countries and have major detrimental effects of socioeconomic development, especially with limited private-sector capital investments in these areas. Finally, this chapter looks at issues that are typically prioritized over internet distribution, such as poverty, hunger, and public healthcare, among others. Although the internet can alleviate many of these issues, the severity is typically so large that more immediate rectification is required and the long-term benefits of the internet is often overlooked and underappreciated. Therefore, as an introductory section, a brief review on Industry 4.0 is presented in the following section.

2.2 Industry 4.0—A Brief Review

In this chapter, several references will be made to the fourth industrial revolution, also known as Industry 4.0. Advances in technology, a shift in the way that business is conducted, and a vast reliance on the IoT all have a significant effect on last mile connectivity. As innovative technologies mature, the cost is inevitably driven lower and developing countries that do not have the financial resources to pioneer these technologies, can implement these technologies and take part in Industry 4.0 developments. A brief overview of what exactly Industry 4.0 entails is presented in this section.

Key terms in defining technologies that shape Industry 4.0 include

- big data,
- cyber-physical systems,
- automation,
- cloud computing, and the
- IoT and industrial IoT (IIoT).

All of these terms drive modern trends of automation, autonomy, and information transfer in manufacturing technologies. There are two primary sectors where solutions provide Industry 4.0 capability, being technology (primarily connectivity and internet access) and organizational structuring. Advances in technology are achieved by educated and skilled individuals and proficient organizations relies on effective and relevant policies supported financially and legally by governments and the private sector. Policies include

- affordability,
- accessibility,
- appropriateness, and
- integration.

These policies must also be dynamic and capable of evolving with the requirements of Industry 4.0 and the sectors where they are implemented. Industry 4.0 is still *under construction* and very few people know exactly what will be the defining technology of the era, or more likely, the combination of defining technologies. Historically, the first three industrial revolutions were defined by specific advances in human adaption, defined by

- The first industrial revolution—spearheaded by the invention of the first efficient steam engine by James Watt, which improved significantly on existing steam engines, improvements and mechanization of blast furnaces, smelting practices, and the invention of the mechanical loom in the textile industry.
- The second industrial revolution (often referred to as globalization)—quick industrial development and growth for pre-existing industries through expansion of long-haul networks of railroads, high-volume and large-scale production, higher efficiency of machinery, telecommunication over long distances through the telegraph, and new energy sources leading to the internal combustion engine.
- The third industrial revolution—defined by nuclear energy, the invention of the transistor leading to computing, and the invention and radical improvements of the microprocessor, driven in large by Moore's law.

All of these shifts in the way that technology changed how business was conducted throughout history has some underlying similarities. Essentially, ways that exclude developing countries from being a part of these revolutions, with developed countries in Europe and North America suppressing the industry by investing large amounts of resources in advancing it, whereas developing countries were not able to match these investments and eventually unable to compete. Industry 4.0 is somewhat different in this respect, and it is essentially an entire industry based on connectivity. This is why last mile internet connectivity in developing countries and rural areas is so important. The European Parliament has listed several main aspects that define Industry 4.0, a convenient breakdown to identify the key aspects. The list include

- real-time adaptive capabilities,
- decentralized technology,
- virtualization,
- interoperability between smart machines and human operators,
- service orientation, and
- expandability.

A general lack of infrastructure in developing countries can be considered an advantage. Developed countries are faced with incompatible infrastructures and legacy issues that must be overcome to allow connectivity that will take Industry 4.0 forward. Albeit these countries have the resources to invest in overcoming these issues, developing countries offer relatively little resistance in embracing disruptive technologies [1]. Furthermore, a historically slow uptake of technology compared to developing countries put developed nations in an ideal position to vault over outdated technologies and implement innovative solutions that have longevity. These innovative solutions are not only confined to the connectivity sector, but also in energy supply. Energy harvesting and renewable energy solutions might be critical due to volatile resources from the grid, but sustainability in the long term can also be guaranteed by clean energy strategies and solutions. However, such major solutions will require strategic partnerships, investments, and a restructuring towards socioeconomic prosperity. Mutual advantages for developing countries investing in modernizing developing nations towards Industry 4.0 should be identified and embraced to provide struggling economies with a head start in a shifting industry. The following section addresses the affordability of mobile broadband, often the only available option for people in rural areas to access the internet due to a severe lack of infrastructure to provide cable internet.

2.3 Affordability

In this section, two primary challenges that hinder the widespread adoption of the internet in developing countries and rural areas are addressed. These challenges are high spectrum licensing fees and a review into the affordability index for populations with limited resources.

2.3.1 Spectrum Licensing Fees

To deliver affordable, widespread, and high-quality mobile broadband services, mobile operators require affordable and predictable access to sufficient radio spectrum [9]. A digital economy is therefore only able to thrive if well-designed spectrum policies are in place. In developing countries, the right spectrum policies can assist in enhancing consumer and social welfare. A negative influence is typically observed if policies seek to maximize state revenues, leading to more expensive mobile services and a reduction in investments in network. The aim, to connect everyone and mitigate the digital divide, a key policy for most governments in developing countries. In countries where a big percentage of the inhabitants resides in rural and remote areas, mobile operators are further encouraged to have affordable and predictable access to sufficient spectrum. In certain emerging markets, the prices of spectrum allocation have been influenced by governmental policies that want to maximize state revenue and operators typically pay similar amounts for spectrum compared to developed countries even in the income of the general population is much lower.

This inevitable trickles down to the consumer since operators may incur financial constraints to protect their return on investment. Current 4G and future 5G technologies require increased amounts of spectrum, and for developing countries to support fast and sustainable access to the mobile sector, spectrum policies must be adjusted consequently. Last mile internet access, in the case of for example millimeter-wave backhaul networks, will also require large spectrum allocations, and to implement such strategies, policies in developing countries and rural areas should be well defined and cost-effective. In the case of last mile technologies such as light fidelity (Li-Fi), spectrum allocation can be relaxed, since the bandwidth used to transmit and receive data is achieved by visible light waves, using light-emitting diodes (LEDs) that already provide lighting. The digital divide in developing countries is already a well-known occurrence, and GSMA [9] represents this data in a well-constructed report. From this report, it is evident that areas such as sub-Saharan Africa, the Middle East and North Africa, Asia Pacific, and Latin America all have below the global average of mobile broadband subscribers, for various reasons (which will become apparent in this chapter). The Commonwealth of Independent States, North America, and Europe, all have above average numbers if subscribers to mobile broadband. Access to radio spectrum is an essential component to delivering mobile services to developing countries and bridging the digital divide [9].

The types of mobile spectrum costs varies and can be categorized. In most cases, an auction for mobile spectrum requires an upfront payment, or through a direct administrative award from a government or regulator to a mobile operator. An annual fee is also payable to cover the running costs of maintaining the spectrum. This annual fee can vary significantly based on the license renewals and the upfront costs involved, payable either before or after the renewal. A government is typically tasked with making three primary considerations when spectrum for mobile services are made, viz

- efficient assignment,
- maximizing consumer welfare, and
- increasing revenue for the state.

These three considerations should be aligned such that the highest value bidders are also the best candidates to realize the highest welfare to the society. Maximizing revenue for the state should not be a top priority for governments, and in developing countries, this is often the case. This is inevitably detrimental to the consumer and the short-term gains for the government can have long-term effects on the socioeconomic welfare of the country. According to GSMA [9], there are several ways how spectrum prices can have an impact on market outcomes. These include:

- If the spectrum prices are inflated, the mobile market can become less profitable. Investors will therefore be likely less attracted to invest in the mobile market, and operators could try to make up lost revenue though increased prices to the consumer.
- If the upfront cost for the mobile infrastructure is high, and repayment cycles are long, unexpected variations in regulatory charges can have a severe effect

on the profitability of the operator, and are difficult to plan for. If repayment cycles are short, charges can be excessive and encumber short-term infrastructure development.

• Inflated spectrum costs would require an operator to finance the charges through debt finance. Repayments and interest accrued will have a negative effect on the cash flow of the mobile operator, again falling to the consumer to provide relief through higher subscription costs.

GSMA [9, 8] examined over 1000 4G spectrum assignments across 102 countries (60 developing and 42 developed) between 2010 and 2017. They found that the average spectrum prices in developing countries has more than doubled in this period. The unit used to normalize spectrum pricing, is US $/MHz/pop, the dollar value of a 1 MHz spectrum allocation per person per year in a selected area. Also noticed in this data, that during this period, specifically between 2013 and 2016, several cases of concerning high prices for spectrum in developing countries have been allocated. The outliers (Jamaica, Iraq, Jordan, Peru, Armenia, Peru, and Sao Tome and Principe) would not necessarily be a cause for concern if the higher price were due to a healthy bidding for the allocation of the spectrum. However, in some circumstances, the prices are driven by government policies through manipulation of the auction design, creating an artificial scarcity of spectrum, or a lack of a spectrum roadmap. An above average administrative assignment cost or auctions with high reserve prices are also among the reasons for these high charges. A cause for concern, presented in the research done by GSMA [9] is that

• between 2010 and 2013, the cost per spectrum as a proportion of income in developing countries was around 2.5 times higher compared to the average in developed countries,
• this value increased to 4 times the average paid by developed countries between 2014 and 2016, and
• reduced back to 2.5 times in 2017.

Furthermore, on average, the spectrum prices as a function of US $ per MHz per gross-domestic product (GDP) (US $/MHz/GDP) in developing countries are three times higher when compared to developed countries. Albeit the GDP of developing countries are on average significantly lower than that of developed nations, it shows that spectrum allocation in these countries can become unaffordable, and these countries will remain a part of the digital divide if allocation policies are not adjusted. However, a finding that contradicts this is the fact that, on average, developing countries pay more than five times more for reserve prices compared to developed countries. The end-result, broadband in developing countries are less affordable than in developed countries, as maintained by the Alliance for Affordable Internet [2].

2.3.2   The Affordability Index

The Alliance for Affordable Internet [2] affordability report is an annual initiative reporting on the progress of policies towards affordable internet globally. The report looks at 58 low-, middle-, and high-income countries to define what changes these countries have implemented to reduce the rates of mobile broadband and to grow access, identifying parts where each country should focus to enable affordable internet access for its population. Adapted from Alliance for Affordable Internet [2], the main recommendations (affordability drivers) for each country are to

- using *public access* (see Chap. 6 of this book) to close the digital divide,
- implementing smart polices that foster market competition,
- have innovative solutions of spectrum usage through transparent policies,
- endorse infrastructure development and the sharing of resources,
- effectively use collective services and access reserves, and
- seeing through effective broadband planning to implementation.

In its opening paragraph, however, the report states that slow progress means that billions of people are still excluded from broadband access, for various reasons outlined in the report. The biggest obstacle: high connectivity cost, especially in Africa. On average, African citizens pay 18% of their monthly income for 1 GB of mobile data. This is close to the 20% housing affordability index prescribed by the National Association of Realtors.[1] The global average for a 1 GB broadband package (prepaid—mobile) as a percentage of the gross national income (GNI) per capita for[2] 2013–2015 is presented in Fig. 2.1.

As shown in Fig. 2.1, the price of a 1 GB broadband plan as a percentage of GNI per capita have decreased on general in all regions from 2013 to 2015 (and assuming the 2018 report would show similar results). The data suggests that Africa and Asia and the Pacific (comprising of multiple developing countries) are still much too high. The report also suggest that a median value of 5% of GNI per capita for a 500 MB data bundle should be the benchmark worldwide.

Focusing on Africa, briefly, Table 2.1 represents the affordability drivers index (ADI) and percentage of GNI per capita for published countries. The ADI assess the extent to which countries have implemented the affordability drivers that can decrease the cost arrangement for broadband. Two main policy groups are considered to determine the ADI score: infrastructure and access.

From Table 2.1, it is noticeable that the price of broadband data in developing countries, specifically looking at countries on the African continent, are high in terms of the GNI per capita. South Africa ranked first on the list, with a 1.48% of GNI per capita, with Zimbabwe having the highest percentage of GNI per capita, 27.93%. The International Telecommunications Union (ITU) price index for 2017 reported

---

[1] A North American trade association that functions as a self-regulatory organization for real estate brokerage (www.nar.realtors).

[2] A value of a country's final income in a year, divided by its population.

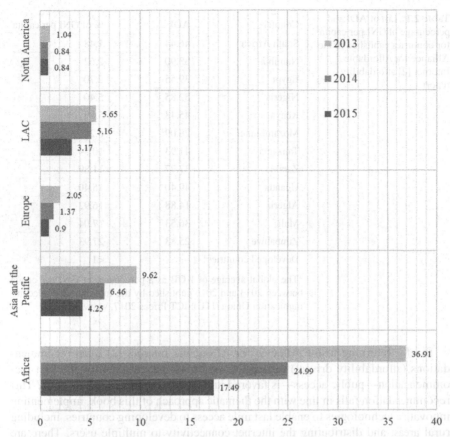

Fig. 2.1 The average price of a 1 GB broadband plan as a percentage of GNI per capita, by region, for 2013–2015. Broadband plans include prepaid and mobile offerings, adapted from Alliance for Affordable Internet [2]

that for 1 GB of broadband data, the average percentage of GNI per capita in the developed world is below 1%.

Somewhat more optimistic from the results, are the ADI scores achieved by each country. The ADI is scored out of a 100, and represents the efforts that a country have made, in terms of infrastructure and access, to make broadband connectivity available in the country at fair prices. Although the ADI scores of the countries in Table 2.1 is typically below 50, it shows that the countries are in fact strategizing and implementing changes locally to improve broadband availability. Nigeria ranked highest in terms of ADI, whereas Algeria presented the lowest score. Worldwide, the Alliance for Affordable Internet [2] affordability report placed Columbia (a non-African country) as the top-ranked country in terms of ADI (also achieving this position in 2016). Columbia achieved an ADI of 72.87 in 2017, with Mexico, Peru, and Malaysia following in the rankings with 71.47, 70.84, and 68.65, respectively.

Table 2.1 List of ADI and percentage of GNI per capita for countries published in the Alliance for Affordable Internet [2] affordability report

Country	ADI	% of GNI per capita
South Africa	46.44	1.48
Namibia	38.90	2.62
Egypt	39.55	2.70
Nigeria	52.85	5.40
Kenya	45.48	5.89
Mozambique	28.09	6.28
Tanzania	41.93	10.54
Zambia	37.77	11.89
Uganda	49.40	15.40
Algeria	14.88	16.92
Mali	36.53	17.04
Zimbabwe	25.83	27.93
Developed countries[a]	–	<1

[a]The global average of 1 GB as a percentage of GNI per capita between 2013 and 2016 published by the International Telecommunication Union (ITU): ICT Prices 2017. www.itu.int

Briefly, returning to the Alliance for Affordable Internet [2] main recommendations (affordability drivers) to improve on broadband availability, the first recommendation—public access—is favored in lieu with the theme of this book. This recommendation falls in line with the thematic approach of this book, implementing innovative technologies to enable last mile access to developing countries, including rural areas, and distributing the internet connectivity to multiple users. There are numerous methodologies and techniques to provide public access, essentially internet in public and open areas, to citizens. This for example consist of the providing internet services on a free of charge or low-cost commercial base. Public access can be granted by the government, civil society groups, businesses, community groups, or private initiatives in a wide range of public areas, such as libraries, community centers, post offices, shopping malls, cafes, and multiple other areas. In view of the quantity of persons living in poverty in developing countries, private internet access is often an impractical objective and public access could be a transition solution. Public access could ensure that people can equally have access to information and contribute to the socioeconomic wellbeing of a country. Furthermore, public access can facilitate educational programs, which would in turn stimulate a greater demand for online services. Governmental policies need to make provision for public internet access but it remains underutilized and its potential unused. The Alliance for Affordable Internet [2] reports marginal improvements in ADI score of average public access policy improvements per income group for the 58 developing countries surveyed, as listed in Table 2.2.

As shown in Table 2.2, the marginal improvements in public access policies for developing countries should be prioritized to increase internet access in these com-

Table 2.2 The average public access policy survey scores by income group, as reported by the Alliance for Affordable Internet [2]

Income group	2014 ADI	2016 ADI	ΔADI
Average: least developed countries	4.10	4.29	+0.19
Low income	4.11	4.60	+0.49
Medium income	4.65	5.22	+0.57
High income	5.74	6.63	+0.89

munities, since private access will still be an unrealistic goal for the near future, since significant infrastructure developments are required and the pricing of these services will have to be lowered drastically. Playing its part in lackluster internet access in developing countries, is a historical lack of infrastructure with limited investments into increasing its footprint. The following section reviews the reasons behind a lack of infrastructure in emerging markets.

2.4 Lack of Infrastructure

In rural communities, connectivity to the internet and other networks are typically supplied by big cellular base stations that have a typical range of approximately 20 km or more, requiring on average, about 1 kW of power, and often times much more, to transmit signals across a specified area. In denser areas, this type of infrastructure is acceptable and cost-effective based on the amount of users it can service. In rural, sparsely populated areas, however, this strategy becomes costly to maintain and operate, and unreliable energy supply leads to frequent downtime and intermittent connectivity. In many developing countries, large cellular towers typically rely on diesel generators for its primary power. This is an extremely expensive solution (compared to coal-powered grid power) and can make up 40–50% of the operating costs of these towers. Solutions such as hybrid power systems (solar, batteries, diesel generators, hydro/wind energy) do exist, but again, are initially expensive and require skilled workers (human resources) to maintain, not always readily available in developing countries where wage rates are often relatively low.

2.4.1 Information Technology Development

Technology transfer infrastructure in developing countries differ substantially with different degrees of robustness across countries [14]. The infrastructure for a specific developing country is primarily a function of its current state of development [14]. An ideal situation is having a network of smaller access nodes able to reach communities in geographically separated areas, targeting these areas specifically and inherently

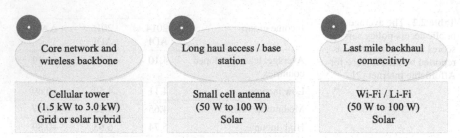

Fig. 2.2 A prospective solution for last mile connectivity in rural communities, albeit requiring substantial investments in infrastructure development

lowering the power consumption and offering the ability to build in redundancy. Such a distributed infrastructure has various advantages, including the power requirements and opportunity of redundancy as mentioned above. Furthermore, it would allow for energy harvesting and renewable energy to supply energy to the distributed nodes to lessen the burden on the already unreliable grid in these areas. A prospective solution for last mile connectivity in rural communities is shown in Fig. 2.2.

As shown in Fig. 2.2, theoretically, a grid-connected cellular tower, requiring between 1.5 and 3.0 kW of power, should serve small cell antennas within the nearby community. Alternatively, and in the long-term a more viable solution, these points can be connected by fiber lines. The small cell antennas in the community can be effectively operated with power requirements of between 50 and 100 W, allowing the use of renewable energy as opposed to relying only on the grid. Multiple instances of these small cells can be distributed as required, and serve local Wi-Fi or Li-Fi hotspots where users can access the internet. In developing countries, however, there is a general absence of fiber optic lines, cellular masts, internet routing services, licensed wireless spectrum, and consistent and dependable electricity; all of which are troublesome in future-proofing these areas for economic development through the availability of broadband internet.

2.4.2 Historical Infrastructure Deficiencies

The primary problem is however, that in many rural communities and more commonplace in developing countries, the infrastructure to facilitate such a strategy is non-existent. Such a solution would require an uninterrupted wired topology among communities, ideally fiber that supplies high-bandwidth internet connectivity to each distributed point, which can then be distributed wirelessly to users using lower-power base stations. Providing multiple fiber links for distances upwards of 20 km is a costly capital investment, and typically overlooked by government as a long-term solution with a high initial investment. Maintaining these links, present high running costs over its lifetime and require skilled workers to identify and repair problems and issues. Therefore, a clear lack of infrastructure, reminiscent from an earlier time where developed countries and urban communities adopted the traditional copper telephone line infrastructures, and rural/developing countries did not, now has a snowball effect since the older solutions can effectively be *replaced* by the high-bandwidth counterparts. It is important for local governments to realize that providing internet to rural communities can (and probably will) become sustainable given enough time. It should be deployed as a means to support economic and social development in developing countries. Specifically, wireless internet in developed countries are mostly associated with mobility and connectivity between homes, businesses, and the internet, as well as a means to facilitate IoT devices designed to provide some functionality that aims to improve on certain tasks, mostly non-crucial and somewhat commercialized. In developing countries, wireless internet is primarily seen as a low-cost solution for broadband last mile internet distribution. Its benefits can only become apparent within the education, health, business, and agriculture sectors once achieved. A historical sufficient and efficient technology infrastructure is critical for supporting future upgrades and data networking for high-bandwidth broadband communications. This includes international data links, availability of medium- to short-range transmissions, as well as high quality local links. A large mean time between failures is ensured by reliable energy sources, efficient maintenance, and quality-ensured installations. Another prominent issue is the cost of installing new infrastructure, a high initial investment, even for developed countries, but without any support from government or investors.

2.4.3 Pricing

Rolling out fiber is a costly exercise, initially. As a result, government backing in many developing countries is still lackluster and the initial investment often outweighs the perceived economic benefits [10]. In the United States of America, several years ago [12], the cost of rolling out buried fiber, as opposed to aerial fiber, was approximately US $75,000 per mile (roughly US $46,600 per km). This includes the cost of the fiber cable, installation, and digging of trenches where the fiber is buried. For aerial fiber (suspended across poles), the cost was somewhat less, at approximately US $50,000 per mile if made ready to use, and US $25,000 for installation only. Since 2011, the prices in the USA have come down significantly, however, in developing countries that import cables and often need to contract out the trenching, the cost still remains in that region. In South Africa for example, a developing country capable of rolling out fiber to a relatively large proportion of its population, the cost in 2015 was still around US $46 per meter [20], equating to roughly US $75,000 per mile. Again, these costs have been reduced with technology advances and locally made equipment and components, but the approximate cost is still high. In Zager [22], a model of the cost of rural fiber deployment is presented, where additional variables such as frost index, wetlands percentage, soils texture, and road intersection frequency increase the complexity of deploying fiber to households.

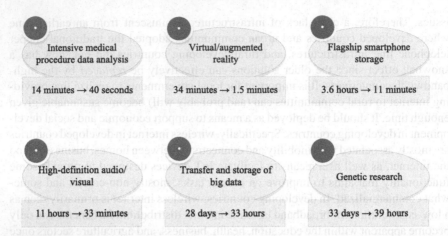

Fig. 2.3 Expected performance improvements from the Gigabit Society 2025 as highlighted by the European commission

For example, in Europe, in 2017 the Fiber to the Home Council of Europe proposed the Gigabit Society 2025 targets. The Commission proposes that all local schools, transport centers, as well as primary providers of public services (and with focus on companies with intensive digital necessities) ought to have access to internet connections with transfer speeds of 1 Gbps (for both upload and download). Furthermore, all European homes, both rural and urban, ought to have access to internet subscriptions with a minimum download speed of 100 Mbps, with the option to increase line speed to 1 Gbps. All urban areas and major road and rail networks should have uninterrupted 5G wireless broadband coverage. The estimated cost for this strategy is €137 billion to provide a complete coverage of fiber-to-the-home (FTTH) across all the EU28 countries and meet the Gigabit Society 2025 targets. The European Commission highlighted certain tasks deemed important to improve upon by this strategy, adapted and presented in Fig. 2.3.

As shown in Fig. 2.3, the tasks highlighted by the European Commission include

- data-intensive medical procedures,
- augmented and virtual reality applications, a sector that will thrive in the near future, both for recreational purposes as well as in the medical field, a probable pillar in the fourth industrial revolution,
- smartphone usage and data transfer, as these devices have become a central part in conducting business in modern times,
- transferring high-definition audio and visual content,
- transfer and storage of big data, and
- genetic research.

Of course, these goals/targets can be considered skewed considering the issues faced in developing countries, but they give an idea of where Industry 4.0 is heading in developed countries. In addition, the total cost of attaining these goals, €137 billion, roughly US $160 billion. This equates to approximately 0.8% of the nominal gross domestic product of Europe. If such an endeavor was to be initiated, in a continent such as Africa, which is mostly developing, the scale of the product would inevitable be much larger. Primarily due to the lack of infrastructure currently. The cost would therefore escalate significantly, where a US $160 billion estimate is already around 5.3% of the GDP of the entire continent, not necessarily a viable option. This percentage is similar to what the United States of America spends on the entire agriculture sector per year. As a comparison on independent country-level, South Africa, a relatively developed country falling under the umbrella of developing countries, has a nominal GDP of US $300 billion, less than double of the Gigabit Society 2025 proposed cost. A country such as Brazil, with a geographical area comparable to that of Europe, would require a GDP percentage of 8.9% for the same project, again notwithstanding its lack of infrastructure currently and escalating cost. Another example, from Zaske [23], is the estimated €2.70 billion (~US $3.31 billion) required by Germany to provide all its citizens with at least 50 Mbps internet access. Comparing the numbers it is evident that developing countries are not able to match infrastructure developments proposed in Europe, with an already established infrastructure (therefore a large percentage is spent on upgrades as opposed to new developments). Emerging markets require significant investments financially, and in skills development locally to obtain internet access for all of its citizens. The following section reviews how monopolies and anticompetitive behavior, especially in developing countries, are causes of low connectivity adoption.

2.5 Monopolies and Anticompetitive Behavior

Historically, telecommunications users have had restricted freedom to choose how their needs should be met [7]. Telecommunications industries only allowed, and to an extent still do, users to rent or buy terminals from the telecommunications companies that owned the user network, owned by the government, by a public, or by a private corporation. The services, standards, and charges are determined by the telecommunications company and the user has very little say in any of these. This has led to monopolistic power of the providing companies over the users, mostly regulated directly or indirectly by the government, eliminating any form of freedom of choice for the user. A similar principle is seen in the manufacturing of telecommunications equipment industry, where limited resources are spent on allowing manufacturers to develop and innovate new technologies until the service provider is able to increase its subscriber base to a satisfactory level. The monopolistic characteristics of this industry was not necessarily born that way, but have since its inception evolved to a state where many of these policies and regulations are necessary to manage the rapid evolution of technology.

Developing countries, especially on the African continent, have made significant strides in the last decade to develop their telecommunications industry; however, a majority have stuck with this monopolistic approach. State-owned enterprises

(SOEs), specifically in telecommunications industry, have retained high prices for consumers and slowed down transformation of several economies. Conversely, a natural monopoly is not necessarily state-owned and governments can sanction anti-trust legislation groups to exploit anti-competitive behavior of public and private establishments. These companies are typically just the first to market a product with widespread applications and are sought after. Many monopolistic SOEs are hindering economic development for a number of reasons, such as being poorly run (rife with corruption) and using outdated, sub-standard infrastructure. For these SOEs to improve their business models and infrastructure, requires large investments of monies that are typically allocated for other sectors, or part of irregular spending and anticompetitive behavior, or not available from financially struggling governments. Since they are already monopolies, with limited to no competition, these entities are able to keep prices high, deter infrastructure development, and hence do not contribute to any economic development within these countries. African countries such as Angola, Cameroon, Ethiopia, and Djibouti have the highest wholesale prices within the telecommunications industry worldwide [16]. Political imbalance, corruption, and financially struggling governments not only prevent telecommunications development and innovation, these issues are difficult to eradicate, as the *damage* is done over many years (often decades), and removing the disarray of the situation(s) would optimistically take just as long, but more likely, much longer. According to Southwood [16], more than half of the 54 sovereign countries in Africa have incumbent state-owned telecommunications enterprises that is either prevailing, or has monopoly liberties, that effectively stifles the development and efficiency of the local market. These sovereign countries are listed in Table 2.3.

A possible and highly debated solution to mitigate monopolies of SOEs in developing countries is privatization of these companies, essentially moving them out of the public (state-owned) sector towards a privately traded company. At times, the government, the previous owner of the investment, remains to exercise a definite extent of control over the trade. This can either be beneficial, or to the detriment of the company. Therefore it is important to decide beforehand the type of privatization that will occur. There are three primary methods of privatization, being

Table 2.3 Sovereign countries in Africa that have incumbent state-owned telecommunications enterprises that are either dominant, or have monopoly privileges [16]

Algeria	Angola	Benin	Burundi	Cameroon	Central Africa Republic
Chad	Comoros	Congo-Brazzaville	Democratic Republic of Congo	Djibouti	Egypt
Equatorial Guinea	Eritrea	Ethiopia	Gambia	Guinea	Guinea-Bissau
Libya	Mali	Mozambique	Namibia	Niger	Sao Tome
Sierra Leone	Swaziland	Tanzania	Zambia	Zimbabwe	

- share issue privatization (SIP) [13],
- asset sale privatization, and
- voucher privatization [16].

SIP is the most common type of privatization and is a process of selling off shares on the stock market, capable of broadening domestic capital markets, boosting liquidity, and potentially growing an economy. In asset sale privatization, an entire organization is sold to a strategic investor, typically through an auction. In developing countries, asset sale privatization is common due to dogmatic and exchange of currency risks discouraging overseas investors. Voucher privatization is the least common, a process where shares of ownership are sold at low prices to all citizens of the country, and typically seen in transition economies. Off course, privatization as a strategy to eliminate monopolistic behavior has its own set of advantages and disadvantages. The concept is often fiercely debated by political powers either supporting government-run services or nationalizing of industries. Advantages of privatization in general include:

- Primarily because of free market competition, it is believed that goods and services can be supplied more proficiently by a private market, as opposed to a public enterprise.
- A privately owned corporation instills accountability on its leaders and its owners and stakeholders mandate performance-based outcomes.
- Profits are at the core of a private organization and a positive revenue stream attracts shareholders. Companies are motivated to entice consumers to use their services or products and to reduce operating cost and irregular expenditure.
- Corruption is typically less prevalent in private corporations as it effects the employees directly, as opposed to being filtered down to the consumer without any accountability for the wrongdoers.
- It inspires innovation and investments in future proofing the business strategy through technology or developing skills of its employees, to ensure longevity of the corporation and an edge over the competitors.
- Privatization encourages competition (leading to employment opportunities), leading to lower prices for the consumer and henceforth socioeconomic benefits for the citizens. It also decreases the inequalities in wealth in a country, where government officials are typically already remunerated with above-average salaries and benefits.

Privatization is a potential solution, aimed at monopolies that hurt the economy of many developing countries; however, the process also has several disadvantages that make it difficult to quantify its effects on a country. Disadvantages of privatization include:

- Not necessarily a disadvantage of privatization, but rather a reason why it would not certainly be in the best interest of the public. In a sense, governments might have some incentive to run an enterprise effectively, because they can harness public support if run successfully. As a method to gain votes for future elections, a government could invest time, skills, and resources into a public entity and make it thrive. Furthermore, a governmental body can also be held accountable, through losing votes, public support, and inevitably ruling power; whereas a private entity can be liquidated and disappear, without compensating consumers and stakeholders at worst case.

- Opponents of privatization deliberate that certain markets such as prisons, basic health care, and basic education should remain segregated from the ruthless and unpredictable private sector.
- Sectors such as the military are also believed to be better suited to governmental rule as opposed to be privatized. The arguments follows that a private organization would use the services of the sector to their advantage and profit, whereas a government would prioritize the safety and wellbeing of an entire nation, regardless of the losses it might experience during unexpected necessities.
- Private companies typically have profit as its main priority, often at the expense of its workforce. Short-term events that fluctuate the markets could be over-analyzed by private companies which would rapidly respond, and in many case, laying off a section of its workforce is the fastest approach to protect investments. Conversely, a governmental approach might be focused on longer terms, with fluctuations in the market often ignored, or mitigated over time, leading to more stability in employment and morality of its employees.
- Privatization's initial benefit to consumers with lower prices for services and goods to entice clientele, might be short-lived, as these industries (especially very successful ones) might inflate prices based on supply and demand, whereas if no control is exerted from governmental bodies, consumers might be trapped into overpaying for services. The organization's initial strategy to achieve success, for example offering the highest quality products, can effectively lead to another monopoly, and with limited incentive for intervention.
- A non-corrupt and properly run government would essentially lose dividends from a successful private organization, which would typically instead be channeled to individuals and not contributing to the social wellbeing of the rest of the population.
- In many circumstances, selling off public assets at inflated prices and replacing civic responsibility with a profit motive, benefits a select few private organizations. The transaction is essentially paid for by the taxpayer.

Privatization is case-specific and involves numerous organizations, governmental bodies, and require public support to be successful. The cases also vary significantly in developing countries when compared to developed countries. Privatized operations are generally more efficient than government operations, especially if there is lackluster support from the government. Giving a private organization free reign to run an operation as they see fit, could also be dangerous and have severe risks for the citizens of the country. A similar argument is valid for the telecommunications industry, and supplying citizens with last mile access, must be handled grounded on the political environment as well as the long-term benefits it could contribute to socioeconomic development. Drastic changes in macro policies and regulations in emerging markets are also required as the methods of doing business and accessing information are changing.

2.6   Macro Policies and Regulatory Challenges

In their goal to align with the Washington Consensus, economic and socioeconomic policies in developing countries have focused mostly on improving the business environment, encouraging macroeconomic stabilization, economic prospects in trade and investments, and the expansion of market forces within domestic economies. Although these policies were initially crafted for rapid economic growth, several critics maintained that such an environment was not focusing on the more pressing issues that developing countries were facing. Challenges that developing countries were, and still are, facing, include

- bottlenecks in infrastructure development as many of these policies were crafted beyond the capabilities of local governments,
- insufficient productive capabilities of the local workforce,
- a significant shortage of skilled workers,
- underdeveloped financial markets, and
- high levels of income inequality.

Innovative and tailored industrialization strategies had to be developed, specifically for the African continent, and are progressively deliberated by business leaders and governmental bodies. The primary reason for this paradigm shift is the realization of what effect connectivity can have on a socioeconomic level. As Industry 4.0 is rapidly approaching, adapting the macro- and micro-policies of governments, private investors, and individuals in developing countries are becoming increasingly important. Several external factors have led to new opportunities to improve and maintain a socioeconomic head start, such as

- enhancements in automation and industrial processes garnered from increasingly digitized industrial processes, allowing individuals to focus on their contribution in technology to enable large-scale manufacturing,
- a global slow-down in global growth after the commodity super cycle, and
- increasing cost of labor in areas such as East Asia, forcing investors and manufacturers to invest in countries that were historically overlooked, advancing the workforce in these areas.

However, for developing nations to take advantage of this global shift in socioeconomic investment, a crucial component needs to be present, broadband connectivity. One of the largest challenges for these nations to overcome is its deficient access to connectivity, on many scales. Policy makers and local governments are urged to shift their focus towards this issue, and overall, last mile connectivity is at the core of potentially solving this problem. Countries must upgrade the ways that business was traditionally conducted and rely more on the internet as a primary channel of communication, data sharing, design, and manufacturing (considering how technologies such as 3D printing has progressed over the last decade). As mentioned in this chapter, the two largest deterrents of ubiquitous broadband connectivity are the cost involved in upgrading, or in many cases, creating a technology infrastructure, and the unreliable energy supplies many of these countries are facing. Connectivity is deemed a major factor in providing a global platform to collaborate with developers, ensure sustainability, improve health services, and socioeconomic advantages such as improving gender equality and education. There is still a vast inequality in the number of people that can access the internet, and if these inequalities are not addressed, progress will be slow.

Bleiberg and West [4] examines economic and political factors that still prevent the majority of the developing world from accessing the internet. The most obvious barriers that Bleiberg and West [4] identifies are

- the lack of infrastructure,
- variation in internet adoption among age groups, especially in senior citizens,
- illiteracy,
- the internet being a fearful technology.

Bleiberg and West [4] furthermore identifies three primary policies that could potentially mitigate the digital divide. These three policies, include

- zero rating services,
- a reduction in connectivity taxes, and
- diversifying the content available on the internet.

Zero-rating services (or referred to as zero-rating) is a technique to provide free internet access under certain circumstances. Defined in various sources such as Galpaya [6], zero-rated data refers to data that does not count towards the allocated data for a user per timeframe, as defined by internet service providers (ISPs). In developed countries, a norm of uncapped data, therefore no significant limits on the amount of data a user consumes per month, is commonplace. However, in most developing countries users typically pay for their data in allocated bundles, specified in megabytes or gigabytes. The cost of this data is typically also expensive and users are not encouraged to use the data on non-essential activities. A zero-rating service, which is essentially a subsidy that provides users to access certain content without using their allocated data. Provide zero-rated services does not only present benefits for the users, but operators and content providers can also benefit through attracting larger user bases which can translate to increased advertising revenue. Additionally, choosing the content that is offered as zero-rating differs when comparing developing and developed countries; ensuring that the most applicable content is zero-rated should lead to socioeconomic benefits.

In general, social media and messaging applications are among the content that is zero-rated. Although providing free social media is not necessarily considered advancing economic development, there are merits in providing such a service. Galpaya [6] refers to research conducted in developing countries within Africa where users use the zero-rated services (social media) to access content such as politics, news, and information about health and medicine. In addition to users having *indirect* access to such information, they are also able to connect with colleagues, friends, and likeminded individuals over the world, which has its merit in a globalized Industry 4.0. Importantly, and an issue that have been under much scrutiny in recent times, is however the production of fake news on these social media sites. Spreading misinformation and propaganda through online social media can have harmful effects and controversy associated by zero-rated services aimed to provide its users with a channel to access information at no cost. Users in developing countries might also have difficulty identifying fake news since they are not necessarily able to access trusted sources (which would incur additional data costs). Market distortion and anticompetitive behavior can also occur on zero-rated services when only providing targeted content that would increase revenue for the operators. As a result, it is required that regulators mandate the content being published on zero-rated services to prevent conspicuous conduct.

Zero-rated services have also ignited a debate between people with radically different views on the matter. A percentage of people argue that zero-rated services violates net neutrality, where users would preferably only access free content and paid content is privileged, and a percentage of people agree that providing some form of (free) content to underprivileged people gives them the opportunity to be more informed, compared to having no access at all. In this debate, factors such as market dominance, user privacy, virtual security, and social equity are addressed and disputed, often leading to policies and regulatory agreements to mandate the primary goal of providing emerging markets with an opportunity to participate in the global economy.

The second strategy in mitigating the digital divide as outlined by Bleiberg and West [4] is a reduction in taxes, specifically concerning connectivity. Essentially, policy makers, primarily governments, should avoid making it more difficult and unnecessarily expensive for investors to participate in advancing terrestrial internet networks by imposing additional taxes or high cost associated with licensing, and therefore challenging access to internet rights across borders. Globalization has already in 2001 lead to many changes in tax policies [3] in terms of

- the level of taxation on connectivity,
- combining taxes,
- developing particular taxes, and
- how specific taxes are administrated and complied to.

As highlighted by Carnahan [5], a well-functioning revenue system is a necessary condition for strong, sustained, and inclusive economic development. Revenue is applied to fund expenditures in physical, social, and administrative infrastructures within a country. It should be used to develop, grow, and expand business and support strong citizen-state relationships that reinforces effective, accountable, and stable governments [5]. Although globalization has led to changes in tax policies, it created new challenges faced by revenue authorities. A prominent issue arising from globalization is base erosion and profit shifting (BEPS) associated with multinational firms, a realization that accumulated impact of domestic corporate tax policies is not adequate to deal with global economic integration. Global services and digital products (e-commerce) can be ordered, shipped, and delivered across multiple borders, involving multiple vendors, and procured by individuals and large organizations alike; a transaction with numerous tax implications if handled correctly. This "shift from a physically oriented commercial environment to a knowledge-based electronic environment poses serious and substantial issues in relation to taxation and tax policies" [11]. E-commerce has the potential to lead to the steady eradication of arbitrators who in the historical sense have been crucial in identifying taxpayers, especially regarding private consumers.

Furthermore, an organization can conduct business in one country; shift the profits to another, effectively lowering the tax implications. This is a strategy commonly followed by developing countries making use of services and labor in developing countries and shifting profits to more mature markets, where lower taxation is enforced. Developing countries with weak revenue administrations are faced with great challenges concerning BEPS and in some circumstances even unaware of the revenue that they are losing in the process [5]. Many developing countries are also rich in resources and offer low-cost labor, a haven for corporations capable (and prepared) to extort them. Developing countries are tasked to identify and manage the effects that e-commerce has on the impact of tax revenue, and depends on factors such as

- the monetary worth of e-commerce trades,
- the strategies employed by national fiscal policy regulations to handle transactions, and
- the efficacy of the internal revenue collection strategies,

to mitigate and minimize revenue losses from taxation changes, which will inevitably increase with Industry 4.0 becoming the global standard.

Finally, according to Bleiberg and West [4], diversification of the content on the internet can mitigate the digital divide by providing content to individuals in their native tongue (considering that most of the developing world does not natively speak English—one of the languages that dominates the internet). Indigenous and under-resourced cultures face a number of obstacles when establishing their languages on the internet [15], especially in "regions and countries with high linguistic diversity, such as Africa, India, and Southeast Asia". Along with English, languages such as French and Spanish typically dominate the internet, whereas for minority languages, it can become difficult to access and comprehend internet content, and more so, the technology and platform used to present the internet to the user. According to the Broadband Commission for Digital Development, an estimated 5% of the world's languages, in terms of the approximate number of languages (between 7100 and 9000), is present on the internet. The Broadband Commission for Digital Development has acknowledged that a major challenge in expanding the internet to the more than 4.2 billion people, who do not have access, is in fact the poor representation of the minority languages in the world. In many instances, even when a population is provided with internet access, for example through a last mile network and a zero-rated service, a local community would not necessarily be able to use the content due to the inherent language barrier. Another major issue relates to internationalized domain names (IDNs). Traditionally, IDNs can programmatically only accept a limited number of characters, including the Latin characters *a* to *z*, numerical digits between 0 and 9, and hyphenation. Even these domain names are troublesome for a large amount of internet users, who do not know and/or cannot understand Latin text. Ensuring that the internet can be accessed without language constraints, a multilingual domain name environment should also be encouraged. A worrying factor globally, specifically prevalent in developing countries, is how corruption on multiple levels can have a long-term detrimental effect on socioeconomic development in a country. These issues and challenges are reviewed in the following section.

2.7   Corruption

In Speiser [17], the 12 utmost corrupt economies globally are listed, defined as the exerting of impact, often by the endowment of monetary favors, to acquire a amenity, based on the financial prudence of 198 countries between August 2012 and August 2014. From this report, it was found that emerging markets in sub-Saharan Africa as well as in the Middle East are detrimentally affected the most due to corrupt relations. The list provided, in ascending order of a predefined scoring system, includes

- South Sudan (Africa),
- Russia (Asia/Europe),
- Myanmar (Asia),
- Libya (Africa),
- Iraq (Asia),
- Equatorial Guinea (Africa),
- Afghanistan (Asia),
- Sudan (Africa),
- Central African Republic (Africa),
- Somalia (Africa),
- North Korea (Asia), and the
- Democratic Republic of Congo (Africa).

A consensus that corruption risks are particularly prevalent in developing countries was agreed on, communicated by Trevor Slack, a legal and regulatory analyst at Verisk Maplecroft, a risk analysis and forecasting company. It was also said that "factors such as weak rule of law and a lack of institutional capacity in developing markets undermine most efforts aimed to combat entrenched systems of patronage. Exposure to corrupt officials and a general reliance on third part agents" are also more prevalent in these areas, according to Slack [17].

In the telecommunications industry, ideally tasked with providing as many citizens as possible with connectivity, corruption is also very prone, especially in emerging markets. The telecommunications industry have experienced rapid growth over the last decade and policies and regulations are constantly evolving. There are high incentives for corruption in this industry, often a result of high licensing fees, exuberant contracting fees, and purchasing of state operators. Strict and coherent management systems are needed to improve on risk assessment, particularly in identifying and dealing with corrupt practices. Anti-corruption policies must be regularly monitored to avoid substantial regulatory and legal risk when entering new markets, with comprehensive and transparent reporting systems, especially for companies bidding for spectrum licenses. In developing countries, these policies and regulatory systems are often not in place, and since corruption is already identified as being highest in these regions, the telecommunications industry experiences high levels of corrupt activity. Furthermore, according to Wickberg [21], with exceeding competition in the telecommunications industry, regulations and regulatory reform are needed to

- supervise the market rivalry and competition through the creation of functional supervisors,
- train the obligatory operators to cope with new competition,
- assign and regulate resources, such as spectrum, in a fair and non-discriminatory means,
- enlarge and improve admittance to telecommunications amenities, and
- endorse and guard customer welfares, including widespread right to use and confidentiality.

Importantly, for regulators to be effective and not be drawn into corrupt practices, these governance bodies should ideally be independent authorities responsible for implementing and administering the regulatory framework and ensuring a form of separation of power from domestic governmental bodies.

From Wickberg [21] and Sutherland [18], corruption in the telecommunications industry typically obstruct people's access to the services by obstructing fair competition and the appropriate regulation of prices, subsequently inflating prices that are severely detached from the cost of providing the services. In many developing countries, corruption is one of the main obstacles to trade in general, among heavy and inefficient bureaucracy and customs regulations. Corruption also has long-term effects on the quality and sustainability of services and infrastructure when telecommunications corporations are less incentivized to train skilled staff or invest in high-quality infrastructure, where bribery and personal relations with contractors oversee these decisions. The burden is frequently felt by the majority of taxpayers in a country, where loss of revenue for the state, because of corrupt practices, often leads to governmental buyouts, essentially funded by the taxpayers.

According to Wickberg [21], there are noticeable ways that corruption affects the telecommunications industry. These effects include

- governmental bodies limiting the number of licenses and concessions to telecommunications operators to favor corrupt companies, effectively stifling competition in a country,
- anti-competitive behavior also leads to inflation of prices to consumers, having significant effects on socioeconomic development,
- supply chain corruption leading to operators opting for the cheapest solutions as opposed high-quality infrastructure, having long-term sustainability issues,
- forms of extortion if corruption eventually (and inevitably) reaches customer services.

Therefore, in order for last mile internet connectivity to be a viable solution, provided to multiple rural areas and specifically in developing countries, corruption must be mitigated to ensure sustainable, cost-effective, and high-quality information links without inflated cost or extorting users. Finally, numerous pressing issues in emerging markets and rural areas require attention and significant investments and international aid to overcome. Often, these issues overshadow the development of the ICT infrastructure in a country, and the following section identifies these issues.

2.8 More Pressing Issues

In this section, several issues and challenges faced by developing countries that typically receive higher priority than providing internet access are identified. The goal of this section is not to re-iterate statistics and reasons why these issues must be addressed first, since it is assumed that the severity and importance of mitigating these problems are understood. The body of knowledge on these pressing issues are well documented and freely available (on the internet), and the reader is encouraged to be informed of the severity, especially in developing countries and rural areas. Last mile internet access, albeit having major socioeconomic benefits that would have long-lasting effects on a society, cannot be prioritized if extreme poverty, hunger, and healthcare epidemics are commonplace in an area. Ignoring these issues are inhumane and they receive as much attention and support to overcome as possible.

A reasonable place to identify and categorize the numerous challenges that large populations of the world face on a daily basis is with the United Nations Sustainable Development Goals for 2030. These goals were adopted on 25 September 2015 to

- wholly end poverty,
- safeguard the earth, and
- warrant affluence for all

as a fragment of a new-fangled sustainable development agenda by the United Nations. Essentially, these goals are optimistic, but attainable if governments, the private economy, and the public culture take part. There are 17 sustainable development goals, adapted and presented briefly in Fig. 2.4.

As shown in Fig. 2.4, the 17 sustainable development goals touch on most challenges and issues that are faced worldwide, whereas these issues are typically amplified in poorer communities and especially in developing countries. This book focuses on developing countries and rural developments, specifically providing these areas with internet access to enhance socioeconomic growth. In lieu with this theme, the following list highlights 14 challenges that cannot be ignored, or under-prioritized, in struggling areas. As a result, a primary drive to provide internet access to all humans as a basic human right might be delayed until the issues presented in this list are addressed.

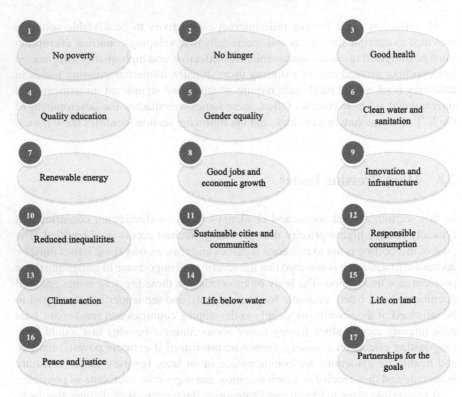

Fig. 2.4 The United Nations sustainable development goals 2030—pressing that must be prioritized to provide improved living conditions to the global population, notwithstanding that factors could be alleviated by last mile access

- Poverty—the condition where basic needs for food, clothing, and shelter are not met. It is also a political issue relating to the allocation and distribution of resources, and reflects the impact of past and present policy choices.
- Hunger—populations and individuals that are not able to intake sufficient amounts or high-in-nutrient food that meets their basic nutritional needs.
- Public healthcare—defending the security and cultivating the health of populations through schooling, policymaking, and study into illness and injury avoidance.
- Public education—facilitating learning, the acquisition of knowledge, skills, values, beliefs, and habits as a basic human right.
- Social and cultural exclusion—a lack of participation in society, at social, economic, political, and cultural levels as a result of the belief system, physical appearance, gender, socioeconomic status, sexual orientation, religion, ethnicity, disability, race, age, and various other factors.
- Clean drinking water—water that is safe to drink, or to prepare food, without any associated risks to the health and wellbeing of a person.

- Electricity—a supply of energy, renewable or non-renewable, enabling an individual to power devices that can be used for heating, lighting, and various other tasks.
- Unemployment—when a person is actively looking for employment but not currently employed because of social or cultural exclusion and/or the economic environment.
- Economic growth—improving the economic, political, and social wellbeing of a nation generally through an increase in aggregate productivity.
- Lack of entrepreneurial infrastructure—provided facilities and services in a geographical area that assists and encourages new ventures and ideas, and grows small- and medium-sized businesses.
- War—armed conflict between two or more groups of people, typically characterized by extreme violence, destruction, mortality, and a significant contributor to slow economic growth.
- Corruption—the misuse of power for public or private gain, typically at the expense of other people, undermining people's faith in governmental and trade and industry systems, organizations, and leaders.
- Agriculture—a combination of population pressure, subsistence agriculture, rural poverty, national resource depletion (such as deforestation), a decline in productivity, urban-biased policies, and gender disparities leading to below-average food production.
- Pollution—poisonous and harmful substances introduced into the environment that have detrimental effects on the health of humans and animals.

In the developing world, especially in urban areas (where an increasing percentage of the global population is residing), many of these challenges are less severe, or even non-existent. As a result, many organizations that are pushing for global internet connectivity are underestimating the current issues that are plaguing certain areas, and should ideally be made aware of these, and provide some form of support to such areas. To its advantage, internet access, and last mile internet solutions, can provide indirect relief for many of these issues. Job creation, educating people, improving skills (both technically and socially), and enabling infrastructure development and a platform for innovation entrepreneurship, can all be addressed in some degree by initiating large-scale projects for last mile internet connectivity in rural areas and developing countries. A reduction in poverty, hunger, improved healthcare, and unemployment will follow suit. Long-term sustainability should be at the core of such endeavors, which requires consistent focus on improving the skills and education of the local population; again, with internet access provided, can become a self-sustaining ecosystem.

2.9 Conclusion

As stated in the introduction, this chapter identifies the challenges that developing countries and rural areas are facing in terms of providing internet access for all. Rather than presenting the reader with all of the statistics on internet connectivity in the developing world, the aim is to enlighten the reader as to why these statistics are true. The challenges covered in this chapter are

- Affordability.
- Lack of infrastructure.
- Monopolies and anticompetitive behavior.
- Macro policies and regulatory challenges.
- Corruption.
- More pressing issues—challenges that do (and should) receive priority over internet access, at least for the short-term.

Bringing affordable internet to developing countries would help it gain access to the global marketplace as well as manage various predicaments, such as health care, education, and sustainability. Enabling people in rural and remote areas to access public services that would otherwise not be obtainable without involving travelling and spending valuable resources. Emerging markets have very different issues and concerns that slow down progress of connectivity infrastructure development, and this chapter identifies and reviews these challenges. Financial barriers, a lack of political support and unstable cultural states are among the primary drivers that hinder ICT growth in many developing countries. Broadband internet is still failing to reach billions of people globally, particularly in the developing world, and stronger efforts by government and international public and private investors are required to mitigate this. This chapter specifically focused on identifying issues that are not necessarily commonplace or well-known in the developed world, and aims to enlighten readers on why these challenges exist (by referring to historical policies) and what are needed to improve access to connectivity.

The following chapter is a technical perspective on the internet. It looks at how the internet is delivered to consumers, and what are needed to distribute the internet physically. Understanding this process is crucial in planning for last mile connectivity to rural areas and serving large portions of users in the developing world.

References

1. Acton A (2018) Industry 4.0: opportunities and challenges for the developing world. Retrieved 21 Aug 2018 from http://www.wwc.co.za
2. Alliance for Affordable Internet (2017) Affordability report. Retrieved 22 Aug 2018 from http://a4ia.org
3. Asher MG, Rajan RS (2001) Globalization and tax systems. Implications for developing countries with particular reference to Southeast Asia. ASEAN Econ Bull 18(1):119–139

4. Bleiberg J, West DM (2015) 3 ways to provide internet access to the developing world. Retrieved 6 Aug 2018 from http://www.brookings.edu
5. Carnahan M (2015) Taxation challenges in developing countries. Asia Pac Policy Stud 2(1):169–182
6. Galpaya H (2017) Zero-rating in emerging economies. Published by the global commission on internet governance. Retrieved 13 Aug 2018 from http://www.cigionline.org
7. Grover KC (1986) Monopoly of telecommunications. In: Foundations of business telecommunications management. Springer, Boston, MA
8. GSMA (2017) Effective spectrum pricing: supporting better quality and more affordable mobile services. Retrieved 22 Aug 2018 from http://www.gsma.com
9. GSMA (2018) Spectrum pricing in developing countries. Evidence to support better and more affordable mobile services. Retrieved 26 Aug 2018 from http://www.gsmaintelligence.com
10. Holt K (2018) FCC funds rural broadband for more than 700000 homes and businesses. Retrieved 29 Aug 2018 from http://www.engadget.com
11. Jones R, Basu S (2002) Taxation of electronic commerce: a developing problem. Int Rev Law Comput Technol 16(1):35–52
12. McGarty TP (2009) Fiber to the home; capital costs and the viability of verizons's FIOS. Retrieved 16 Aug 2018 from http://www.researchgate.net
13. Megginson WL, Nash RC, Netter JM, Schwartz AL (2000) The long-run return to investors in share issue privatization. Financ Manage 29(1):67–77
14. Sadowsky G (1993) Network connectivity for developing countries. Commun ACM 36(8):42–65
15. Schwab K (2015) The internet isn't available in most languages. Retrieved 13 Aug 2018 from http://www.theatlantic.com
16. Southwood R (2014) Top 5 telco monopolies hurting Africa. Retrieved 22 Aug 2018 from http://www.businesstech.co.za
17. Speiser M (2015) The 12 most corrupt countries in the world. Retrieved 13 Aug 2018 from http://www.businessinsider.com
18. Sutherland E (2011) Corruption in telecommunications: problems and remedies. Info 13(5):4–19
19. UNCTAD (2018) Adapting industrial policies to a digital world for economic diversification and structural transformation. In: Trade and development commission multi-year expert meeting on enhancing the enabling economic environment at all levels in support of inclusive and sustainable development, and the promotion of economic integration and cooperation. Second session, Geneva, 19–20 March 2018
20. Vermeulen J (2015) How much FTTH really costs to roll out. Retrieved 6 Aug 2018 from http://mybroadband.co.za
21. Wickberg S (2014) U4 expert answer: overview of corruption in the telecommunications sector. Transparency International. Retrieved 9 Aug 2018 from http://www.u4.no
22. Zager M (2011) Modeling the cost of rural fiber deployment: an important new study of independent telco FTTH deployments provides a cost model that both providers and policymakers can use. Broadband Properties, pp 106–108
23. Zaske S (2015) As Germany prepares to spend E2.7bn on broadband, vectoring fight breaks out. Retrieved 6 Aug 2018 from http://www.zdnet.com

Chapter 3
Signal Propagation and Networking Fundamentals Required in Last Mile Connectivity Planning

Abstract The innovation towards last mile solutions that are capable of distributing internet access in emerging countries could, and should, emanate from the people facing these issues daily. The fundamental principles of transferring information from one point to another should be understood in order to challenge and improve current technologies that successfully serve the developed world. Electrical and optical signal transmission, as well as radio wave propagation, are at the core of transferring information over a medium. Attenuation, an inevitable physical property of any signal, determines the feasibility of a specific technology in a particular environment. Rural and underdeveloped areas often lack the infrastructure that delivers connectivity from the core network to communities. These communities are typically situated geographically distant from backhaul networks. To effectively plan and design last mile solutions to serve these areas, signal propagation through various media should be researched in order to determine the solutions that compliment the needs of each individual community. Once connectivity has been established, the fundamental principles of networking are also required to design efficient architectures that are capable of delivering the internet to end-user devices. This chapter reviews all these fundamental principles and serves as a reference for researchers in emerging markets that are committed to bringing internet connectivity to disconnected areas. The chapter provides the theoretical knowledge and research methodologies required when analyzing feasible technologies for solutions that are area-specific.

3.1 Introduction

Emerging technological disruption and innovations in last mile internet access are forecasts of growing service deliveries, evolving customer expectations, increased competition, and the growing necessity to provide emerging markets with internet access to allow them to take part in the fourth industrial revolution. Business models are evolving to take advantage of high-speed, always-available internet access to provide a new range of services, with new frameworks and policies being developed to expand on current solutions and develop innovative modern solutions in various sectors. Advanced computer algorithms and analytics are improving last mile service

© Springer Nature Switzerland AG 2019   71
W. Lambrechts and S. Sinha, *Last Mile Internet Access for Emerging Economies*,
Lecture Notes in Networks and Systems 77,
https://doi.org/10.1007/978-3-030-20957-5_3

delivery and are opening doors for establishing new entrants in the digital space, importantly also allowing developing nations and rural areas to participate.

Current last mile technologies are in many cases sufficient for delivering networked and internet access to multiple users, but its distribution remains bound to infrastructures that allow for easy and cost-effective high-volume expansion. Technologies such as 2G/3G/4G, copper-based Ethernet, and fiber networks are among the most common implementations of last mile technologies, however all of these are overlooking factors that are assumed to be commonplace in many developed countries:

- access and availability to relatively expensive end-user devices,
- efficient and reliable energy from the grid,
- network coverage,
- low-levels of security for both physical equipment and for data/information,
- skilled workers that install and maintain infrastructures,
- financial backing for government or the private sector to fund infrastructure development and expansion, and
- access to current-generation technologies to deliver last mile access.

However, many of these factors can be shown to be unreliable or completely unavailable in rural areas or in developing countries, and last mile access requires innovative solutions and extensive planning to implement sustainable access for large numbers of users that could potentially participate in the fourth industrial revolution.

Current last-mile technologies are broadly divided into three categories, namely;

- copper-based wired technology,
- fiber-based wired technology, and
- radio-wave-based wireless technology.

Each of these technologies have their own set of advantages and disadvantages, being it from a cost perspective, maximum throughput, distance it can cover, security characteristics, or geographical feasibility. Current generation technologies as well as future technologies all operate on similar principles in terms of propagation models of the transmitted signal and throughput limitations. These principles include, for copper-based wired technologies, the propagation models for electrical signals through conductors, and for radio-waves, propagation models of electromagnetic signals through air (or any other medium). In terms of optical communication, similar propagation models compared to radio waves can be adapted, since light waves are essentially electromagnetic waves at very high frequencies. In this chapter the essential considerations of propagation models are reviewed, and factors that typically attenuate a signal sent from a transmitter to a receiver, identified. These factors, along with the propagation models, are expanded on in the rest of this book when discussing the characteristics of future technologies such as light fidelity (Li-Fi) and millimeter-wave (mm-wave) backhaul technology.

Furthermore, this chapter also discusses the crucial building blocks of a network, capable of delivering internet access within the last mile to the end-user. Any network, regardless of its physical size and number of users served, require specialized network hardware and software to deliver data and information from the internet to the end-users. These building blocks must be identified in order to understand and plan for developing infrastructures capable of delivering last mile access. In this chapter, these building blocks are identified, with specific reference to current last mile technologies, and will be referenced to in consequent chapters of this book. They serve as a reference for understanding on what is required in a typical network, and how/where future technologies can play a role in replacing current implementations that lack physical expansion in developing countries and in rural communities.

The following section is dedicated to reviewing the expected losses and the mechanisms responsible for signal attenuation in copper-based wired technologies, such as electrical cables. The factors that influence signal integrity are highlighted and the following section presents the user with the tools needed to plan for potential last mile delivery solutions.

3.2 Electrical Signal Attenuation

An electrical cable has an inherent power loss because of its resistivity and the current flowing through the cable at a given time. These losses must be accounted for in power cables, but also when used to transmit any electrical signal such as data between two points. The frequency of an alternating current (AC) signal through an electrical cable also has an effect on the losses experienced during transmission. One of the main culprits of direct current (DC) losses is copper losses, also referred to as Joule heating or I^2R losses. As is apparent from the name I^2R losses, this type of attenuation in the electrical signal is a function of the current and the resistance of the wire. It can be shown below that this effectively means that the losses are also dependent on the diameter of the wire as well as its length. Therefore, the length of any electrical cable must be kept as short as possible, and with respect to last mile access, wiring is a substantial contributing factor leading to the quality of the internet experience of the end-user.

Copper losses are determined either by determining the resistance of the wire or by the voltage drop experienced between the source and the destination, effectively resulting in the same answer. If using the voltage drop strategy, it is key to consider that the resistance of the wire, R_{cable}, is determined by the relationship between the current and the voltage, such that

$$R_{cable} = \frac{V_S - V_D}{I} \qquad (3.1)$$

where V_S and V_D are the voltages at the source (transmitter) and the destination (receiver), respectively, given in volts. To determine the copper losses, in terms of power in watts, as a result of the resistance of the cable, the following approach is used, where

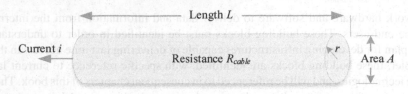

Fig. 3.1 A single conductor such as an electrical wire and the parameters used to determine its electrical resistance theoretically, as opposed to measuring the voltage drop between the transmitter and receiver

$$P_{loss} = I^2 R_{cable} \tag{3.2}$$

and P_{loss} represents the (power) losses in the wire. The main problem with this approach is that to know what the voltage drop in the cable is, it is typically measured practically. Therefore, the cable must first be installed after it can be measured, or at least manufactured to the specific length and measured in a laboratory environment. This is off course not a practical approach and eliminates the capabilities of simulating various cable lengths to determine the expected losses. It is therefore recommended to determine the resistance of the cable by taking into account its resistivity (and conductivity) which are functions of its geometry and the material used. Figure 3.1 is a representation of a wire or cable and the parameters used to determine its electrical resistance.

In Fig. 3.1, if a uniform cross-section and a single material of construction is assumed, the resistance R of the material is determined by

$$R_{cable} = \rho \frac{l}{A} \tag{3.3}$$

where ρ is the resistivity of the material (such as copper), l is the length of the cable and A is the cross-sectional area of the wire. The resistivity of the material is dependent on the material itself, and can be found from various online sources. Some of the commonly-used materials and their electrical characteristics are given in Table 3.1. The cross-sectional area of the wire is determined by considering its geometry, which is typically circular, and is determined by

$$A = \pi r^2 \tag{3.4}$$

where r is the radius of the circular wire in Fig. 3.1. As mentioned, Table 3.1 gives the resistivity, and the conductivity (which is the inverse of resistivity) of some commonly-used materials. Conductivity δ (in Siemens per meter) is sometimes used as an alternative way of expressing the conductive capabilities of a material since its unit range is typically easier to use, as seen in Table 3.1.

From Table 3.1, the most used material for electrical cables, copper, has a resistivity of 1.68×10^{-8} Ω-m relating to a conductivity of 5.95 S/m. Silver, albeit too

Table 3.1 Electrical resistivity and conductivity of materials that can be used to conduct electrical signals, values specific to 293 K

Material	Resistivity (Ω-m)	Conductivity (S/m)
Copper (Cu)	1.68×10^{-8}	5.95
Annealed copper (Cu)	1.72×10^{-8}	5.81
Aluminum (Al)	2.65×10^{-8}	3.77
Platinum (Pt)	10.6×10^{-8}	0.943
Nichrome (Ni, Fe, Cr alloy)	100×10^{-8}	0.0004
Carbon (graphite)	3.60×10^{-5}	–
Silver (Ag)	1.59×10^{-8}	6.29

expensive to implement as a standard material, has a lower resistivity of 1.59×10^{-8} Ω-m and therefore conductivity of 6.29 S/m. Since the resistance of a cable increases linearly with its length, several techniques exist to limit the amount of attenuation experienced along its part. Alternative mitigations include

- increasing to power deliver from the source (will lead to higher electricity cost and more expensive equipment),
- increasing the area of the wire (total weight of the cable will increase significantly and drive up the overall cost of the cable and its installation, especially for long distance installations),
- decreasing its length (a viable alternative if practically feasible),
- adding repeater/gain stages along its path (increasing the cost of the installation with added complexity and equipment),
- or using a different material (using materials other than copper could have advantages through lower resistivity, but will drive up the cost of the cable significantly).

There are however trade-offs to each of these strategies as mentioned in the list above, and the procedure followed depends on the application and the environment where it is implemented.

For AC circuits, additional losses are inherent when transmitting a signal between two nodes. During AC transmission, the ratios between the voltage and the current flowing through the conductor is not only a function of its magnitude, as is the case with DC signals, but also of its phase. A conductor that is not ideal (therefore any practical conductor) will have some amount of inductance and capacitance associated with it. As a result, the *time* that the voltage and the current reaches a maximum, might be different, compared to for example an ideal resistive component. The length and the cross-sectional area, as well is the material, geometry, and surroundings of an electrical cable all have an impact on the AC losses experienced. Furthermore, the electrical characteristics of the signal such as its frequency (leading to the skin-effect) additionally factor into the equations; this leads to relatively complex equations to determine the amount of losses experienced in conductors. To determine AC losses, the signal components are typically divided into real and imaginary components (as

Fig. 3.2 Ohm's law
representing a complex
resistive component in an
AC circuit

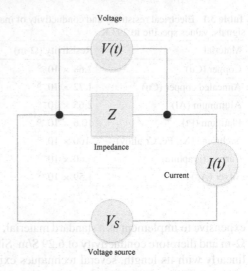

a result of the inductive and capacitive behavior and the magnitude/phase representations).

To derive the electrical impedance (the resistance in a AC circuit), Fig. 3.2 represents Ohm's law as a function of the signal source and a non-ideal resistive component; this resistive component could represent for example the electrical resistance
of a cable used to distribute network connectivity between two nodes.

As shown in Fig. 3.2, the signal source is a AC signal and is denoted by V_S, where
$I(t)$ is the current in the system is represented by, the voltage across the load (or the
total impedance of the cable) is represented by $V(t)$, and the impedance component
is given by Z. The total impedance is therefore given by

$$Z = R + jX \qquad (3.5)$$

where R is the real component of the impedance (therefore the traditional resistance)
and X is the inductance/reactance of the component. The inductance (X_L) is a function of the inductive properties (imaginary) and reactance (X_C) is a function of the
capacitive properties. The inductance is determined by

$$X_L = 2\pi f L \qquad (3.6)$$

where f is the frequency of the AC signal in Hz and L is the inductance of the wire
(typically a parasitic component since it should ideally be zero). The reactance is
determined by

$$X_C = \frac{1}{2\pi f C} \qquad (3.7)$$

where C is the (parasitic) capacitance of the wire. Both the inductance and reactance is measured in Ω. As a result, whenever a time-dependent (frequency-dependent) signal is passed through such a component, the electromagnetic field opposes the rise and fall of the current flowing through the component. If the component is purely inductive or purely reactive (typically not possible in practical components, not even in superconductors) there will be no real resistance R. The total impedance Z of the component would therefore be a function of the real resistance and the imaginary component, such that

$$Z = \sqrt{R^2 + X^2} \qquad (3.8)$$

where X represents either the inductance or reactive component. To determine the phase angle of the signal as a result of the imaginary components, the cosine function can be implemented such that

$$\cos^{-1}\phi = \frac{R}{Z} \qquad (3.9)$$

where Z is comprised of the sum of the inductive and reactive components. Also noticeable from the equations for inductance and reactance, is the frequency component. In a purely inductive component, the inductance increases with an increasing frequency and can become extremely large for frequencies in the high-GHz range. As a result, if the inductance X_L reaches a high enough value, the current that flows through the device becomes zero—therefore any current flow is opposed. Conversely, reactance decreases with increasing frequency and at very high frequencies, a purely capacitive component will act as an ideal conductor and not oppose the flow of current. At lower frequencies, however, the capacitive reactance starts opposing the flow of current and at frequencies close to 0 Hz (therefore DC), it will completely oppose current flow and act as a perfect isolator. For both these reasons, it is an important design consideration in medium- to long distance transmissions to carefully determine the current flowing capabilities through a cable; whether used to transmit static power of frequency-dependent signals, which is typically the case when transmitting digital data (bits) at specific rates through an electrical cable.

To mitigate the losses experienced by copper (or any other) cables completely, another technology is required to send information among nodes, especially when the bit rates become high and the imaginary components that oppose the flow of current become large. One such solution, albeit with its own set of advantages and disadvantages, is using optical fibers. The following section reviews the typical losses that can be experienced when transmitting data through an optical fiber. Last mile access through optical fiber have been gaining traction in recent years, but is still a relatively expensive alternative to copper cables; but is used in conjunction with other technologies to drive down total cost. As the title of this chapter suggests, the current technologies used to achieve last mile access is discussed here, and future technologies such as Li-Fi and mm-wave backhaul technologies are discussed in the following chapter. It is also important to note that the principles reviewed in this chapter is also relevant for future-generation technologies, but typically to a larger extend as technologies are pushed to the limits. Therefore, radio wave propagation models are also relevant in mm-wave backhaul technology, and attenuation in optical fibers can be related to light-based communication in Li-Fi.

3.3 Optical Signal Attenuation

3.3.1 Types (Modes) of Optical Fiber

As a prelude to optical signal attenuation within an optical fiber, the two primary types of optical cables are briefly reviewed. These types refer to single-mode and multi-mode cables, and the importance of understanding the difference becomes apparent in link budget calculations and influences the planning of last mile solutions. Multi-mode cables are further categorized as step-index and graded-index cables. Essentially, the difference between single-mode and multi-mode cables are

- the diameter of the core,
- the type of light source (in terms of its wavelength), and
- the type of modulation.

Figure 3.3 represents the difference between a single-mode and a multi-mode optical fiber cable.

The diameter of the core of single-mode cables are smaller compared to multi-mode cables and only allows a single mode of optical light to broadcast over the cable. As a result, in single-mode cables, the amount of light reflections during propagation decreases, leading to lower attenuation factors and importantly, larger distances that can be traversed by the light. A typical single-mode cable diameter is 9/125 μm (core to cladding ratio). The larger diameter-core in multi-mode cables allows multiple modes of light to propagate, increasing the bandwidth of the cable, but at a cost of an increase in light reflection leading to more attenuation (due to high dispersion) per unit length. The core to cladding ratio of multi-mode fiber is typically rated as 50/125 and 62.5/125 μm. A single-mode fiber, with its relatively narrow diameter, typically allows single modes of either 1310 nm light waves, or 1550 nm waves to propagate at any given time. These light sources must be able to provide a light wave of very narrow spectral width, and since there is no overlapping light among multiple light sources, the attenuation and therefore the distance that the light can propagate is larger than that of multi-mode cables. The increased bandwidth over short distances provided by multi-mode cables, typically make use of light sources at 850 or 1300 nm.

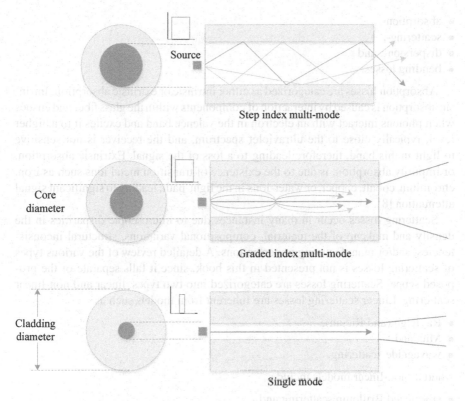

Fig. 3.3 A representation of the difference between a single-mode and a multi-mode optical fiber cable

3.3.2 Signal Loss Mechanisms

Attenuation of an optical signal essentially means the attenuation or losses of energy, in the form of light, as it passes from the transmitting side of the cable to the receiving side, therefore between the source and the receiver. Optical signal loss is also referred to as signal loss or fiber loss. The attenuation experienced in an enclosed system essentially determines the number of repeaters and the distance between them, placed between the transmitter and the receiver. As with traditional copper cables as well as with radio waves, the attenuation in an optic fiber is directly proportionate to the dimensions of the cable (width and length). Attenuation α is simply designated as the fraction between the input and the output power, such that

$$\alpha = 10 \times \log_{10} \frac{P_{in}}{P_{out}} \tag{3.10}$$

where P_{in} and P_{out} are the input and output power, respectively. The various types of losses that are experienced in a fiber optic cable are

- absorption-,
- scattering-,
- dispersion-, and
- bending losses.

Absorption losses are categorized as either intrinsic or extrinsic absorption. Intrinsic absorption is caused by interaction of components within the glass fiber and ensues when photons interact with an electron in the valence band and excites it to a higher level, typically close to the ultraviolet spectrum, and the receiver is not sensitive to light in this band, therefore leading to a loss of the signal. Extrinsic absorption, or impurity absorption, is due to the existence of transition metal ions such as iron, chromium, cobalt, copper, or water-ions in the light-path, leading to significant signal attenuation [8].

Scattering losses occur in many instances due to microscopic disparities in the density and makeup of the material, compositional variations, structural inconsistencies, and/or manufacturing imperfections. A detailed review of the various types of scattering losses is not presented in this book, since it falls separate of the proposed scope. Scattering losses are categorized into two types, linear and non-linear scattering. Linear scattering losses are inherent from models such as

- Rayleigh (and Rician),
- Mie, and
- waveguide scattering,

whereas non-linear models include

- stimulated Brillouin scattering and
- stimulated Raman scattering

and various online sources can be referenced for in-depth analysis on these scattering losses. As reference, if a received signal comprises of several reflective waves as well as a major line of sight (LoS) element, the envelope amplitude from small-scale fading has a Rician probability density function and the non-fading constituent is termed the specular component [8]. If the absolute size of the specular constituent approaches a zero-value, the probability density function leans towards a Rayleigh distribution (also termed the random or scatter or diffuse component [8]), and this is expressed as

$$p(r) = \frac{r}{\sigma^2} \exp\left[-\frac{r^2}{2\sigma^2}\right], \quad r \geq 0 \tag{3.11}$$

and

$$p(r) = 0, \quad r \geq 0$$

where r is the envelope amplitude of the received signal, and $2\sigma^2$ is the pre-detection mean power of the multipath signal. The Rayleigh function, with no specular component, therefore represents the "*worst-case of fading per mean received signal power*" [8]. To determine attenuation and delay rates of mm-wave signals due to fog or clouds, calculations are typically derivative of a complex refractivity grounded on the Rayleigh absorption approximation of Mie's scattering theory [6]. Mie scattering occurs when the molecules or particles that cause the scattering are larger than the wavelength of the signal, whereas Rayleigh scattering primarily calculates scattering from particles smaller than one 10th the signal wavelength.

Nonlinearities in a medium of transmission results in Brillouin scattering, specifically related to the nonlinearity of acoustic phonons [2]. During emission of light, incident photons can be transformed to scattered photons, comprised of a somewhat subordinate energy, and typically propagating in the backwards direction, as well as an emitted phonon. If a certain threshold of power is reached during transmission of a light beam, stimulated Brillouin scattering is capable of reflecting a significant part of the incident ray and contains a large component of nonlinear gain in the optical domain. The frequency of the ray this is being reflected is also subordinate compared to the incident ray and the change in frequencies corresponds to the rate of the released phonon. This is denoted to as the Brillouin frequency shift (v_B), and can be calculated by

$$v_B = \frac{2n v_a}{\lambda} \tag{3.12}$$

where n is the refractive index, v_a is the acoustic velocity, and λ is the wavelength, and the factor depends on the temperature and pressure (exploited primarily in optic fibers).

Finally, Raman scattering is created by a nonlinear response caused by pulsations in the crystal (glass) lattice of the optical medium. It is therefore due to a nonlinear reaction of a transparent medium to the optical strength of light. Raman scattering can transfer much of the energy of a pulse into a wavelength range where laser amplification does not occur, a detrimental effect of Raman scattering.

Dispersion losses are primarily due to light becoming distorted as it travels along a path in an optical fiber, and is categorized as inter-modal or intra-modal dispersion. A consequence of propagation delay differences among modes in a multi-mode fiber is inter-modal dispersion, whereas intra-modal dispersion occurs in single-mode operation, leading to pulse spreading from material (spectral/chromatic dispersion from refractive index differences of the core as a function of the wavelength) or waveguide dispersion.

Bending losses are divided into two primary categories, being either macroscopic of microscopic bending losses. Macroscopic bending occurs when the complete fiber undergoes a significant bend causing some modes not to undergo reflection and leading to losses in the outer layer. If the cladding or the core of the medium undergo small bends at the surface, causing reflections of light at angles that do not allow for further reflection of modes within the cable, the phenomena is called microscopic bending.

3.3.3 Signal Loss Calculations

To calculate the signal loss in an optical fiber, there are typically two different calculations required to determine the total estimated losses in the medium. Each of these calculations takes into account that there measured values which are therefore accepted as known values for various collections of measurements. These assumptions include

- maximum signal loss and
- maximum fiber distance if the link budget and the loss variables are known or predetermined.

In a predetermined interval of fiber, the highest value of signal loss is purely the summation of all worst case values within each section, and these values depend on experimentally determined values. These variables include

- the fiber attenuation per kilometer,
- attenuation per splice multiplied by the number of splices,
- attenuation per connector multiplied by the number of connectors, and
- a predetermined safety margin to account for unknown tolerances.

Once the signal attenuation through the cable is determined, the maximum signal strength must also be estimated, to avoid data loss problems or overdriving the receiver. Secondly, the maximum fiber distance should be calculated and this requires knowing the

- launch power and the
- receive sensitivity

of the specific fiber. These values are typically specified by the manufacturer of the fiber and can be found in the datasheet or other documentation that are provided from the vendor upon purchase of the fiber optic cable. If the maximum distance is calculated, the *lowest* launch power must be taken in order to compute the worst-case scenario. To ensure that the receiver is not overdriven, another calculation using the highest launch power should be performed. The largest distance calculation is determined by first determining the power budget, a simple calculation of subtracting the receiver sensitivity from the launch power, typically expressed in dBm. In order to convert from watts (commonly specified in milliwatts (mW)) to dBm, the following equation can be used:

$$dBm = 10 \times \log_{10} mW \tag{3.13}$$

and to convert from dBm to mW,

$$mW = 10^{\left(\frac{dBm}{10}\right)} \tag{3.14}$$

Table 3.2 Fiber loss table

Mode/wavelength	Core diameter (μm)	Attenuation (dB/km)	Attenuation per splice (dB)	Attenuation per connector (dB)	Modal bandwidth (MHz-km)
MM, 850 nm	50	2.40	0.1	0.75	500
MM, 850 nm	62.5/125	3.00	0.1	0.75	200
MM, 1300 nm	50	0.70	0.1	0.75	500
MM, 1300 nm	62.5/125	0.75	0.1	0.75	500
SM, 1310 nm	9	0.35	0.01	0.75	Not applicable
SM, 1550 nm	9	0.22	0.01	0.75	Not applicable

Legend MM = multi-mode, SM = single-mode

with these equations being helpful if a vendor specifies its launch power and/or receiver sensitivity in mW as opposed to dBm. Once the power budget is determined, the individual loss-components can be subtracted and the net power budget is then used to determine the maximum distance. If, for example, the net power budget is determined to be 10 dB (power budget—splice attenuation—connector attenuation—safety margin), and the cable-loss is 2.5 dB/km, an absolute maximum distance of 4 km is recommended. In Table 3.2, a list of fiber losses with respect to the mode of operation and the wavelength of the source is presented, typically used to calculate the maximum distance and link budget of a fiber network.

As shown in Table 3.2, the attenuation per km of a multi-mode 850 nm fiber is the highest, at 2.40 dB/km. The lowest attenuation is achieved in single-mode operation (at 1550 nm) at only 0.22 dB/km (with attenuation figures of 0.14 dB/km reached in Tamura et al. [10] for a 1550 nm fiber).

Finally, in terms of signal propagation models, the basic principles of radio wave propagation are reviewed in the following section. Again, its importance stems from the fact that these models can be referenced to determine the expected effects of implementing radio signals in various environments. Aspects such as the span between the source and the receiver and typical obstructions that hinder signal propagation, and the effect this has on signal integrity, are presented here and should be used as a baseline when planning for last mile delivery using radio waves (wireless technology).

3.4 Radio Wave Propagation Models

Radio wave propagation models are often used to characterize signal attenuation (path losses) or the impulse response of a signal. These models are typically specific to a site of environment and there are broad generalizations that can be implemented when developing these models. In this chapter, various wireless technologies that enable last mile access are analyzed. It is therefore apt that a theoretical introduction into expected signal integrity based on factors such as frequency, distance and obstructions is presented before discussing these wireless alternatives. Inherent trade-offs such as between the bandwidth capabilities and the transmission distance of wireless signals have an impact on the practicality and feasibility of a proposed technology.

In general, propagation models are defined by two primary categories, namely deterministic and statistical [7]. Deterministic models require large amounts of measured data with respect to geometry, obstruction materials and its effects on the signal integrity, human interaction with the signals and in many circumstances also environmental conditions such as weather patterns. In reality, obtaining adequate data to effectively model a specific environment can be intensive, time consuming and not always practical.

Statistical models aim to alleviate the limitations of deterministic models and only study the physical characteristics of the communication network and do not rely on comprehensive information of the geometrical or electromagnetic features the objects and materials within the circumstances. These models are built on the statistical classification of the received signal [7] typically from previous channel transmissions and data.

As described in Perez-Vega et al. [7], the characterization of a propagation model is typically performed by considering a discrete linear filter, regardless of the type of propagation model (deterministic or statistical) and defining a propagation channel. The output $y(t)$ of this discrete linear filter with respect to a stimulated time-dependent input, such as $x(t)$, is determined by

$$y(t) = \sum_{k=1}^{N} a_k x(t - \tau_k) e^{j\theta_k} \tag{3.15}$$

where the path is defined by k, τ_k is the signal delay in seconds, a_k is the amplitude of the signal, θ_k is its phase, and N is the number of paths that the electromagnetic signal follows between the source and the receiving entity. N is therefore dependent on the physical environment and the signal generally not only follows a single LoS path, but also has multiple alternative paths to reach the receiver (for instance reflected from a ceiling in an indoor environment). A typical input stimulation is the unit impulse $\delta(t)$, in which case the output $h(t)$ of the linear filter is described by

$$h(t) = \sum_{k=1}^{N} a_k \delta(t - \tau_k) e^{j\theta_k} \tag{3.16}$$

and to define the impulse response, the amplitude, phase and time delay of the signal must be known [7]. It is assumed that θ_k and a_k follow a uniform distribution for a Rayleigh fading channel, and the time delay can be modeled by a modified Poisson process. The detailed modeling approach is not presented in this chapter, as it falls outside the scope of the book. The importance of this discussion, is to understand which parameters and characteristics of a signal determine its behavior in a typical environment. The complexity of propagation models can vary significantly and can be modified and perfected to give good approximations of signal integrity and required power budget and therefore predict its coverage area. However, these approximations are still site-specific and a generalized representation of the propagation characteristics is often sufficient to define the coverage area and bandwidth capabilities of a wireless implementation. In context of last mile access, bandwidth often receives priority over coverage area, especially if costly modifications to an infrastructure would allow for only a few meters of gained coverage. Additionally, the coverage area could be pre-defined and indicated to prospective users, as is the case in many developing countries with internet-enable zones. Serving as many clients with high-speed network access is therefore often prioritized in lieu of last mile access. Furthermore, the high inverse proportionality between distance and operating frequency (therefore coverage area and bandwidth)—as will become apparent in the following paragraphs, is often more difficult to overcome than to keep a constant coverage area and increasing bandwidth. Line-of-sight (LoS) and objects that attenuate the primary signal also have a significant effect on designing for last mile access and their theoretical consequences are also discussed in this chapter.

A simplified representation of the primary factors that influence the integrity of a wireless signal are shown in Fig. 3.4. These factors are typically used to estimate coverage and throughput of a wireless signal, and are dependent on the frequency of the signal and the distance between the source and the receiving entity.

As shown in Fig. 3.4, the key factors that result in attenuation of electromagnetic wave propagation are

- reflection,
- diffraction,
- and scattering.

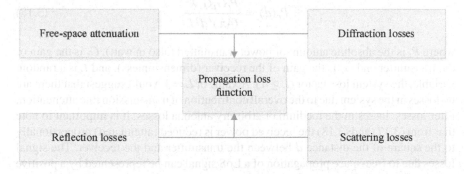

Fig. 3.4 Typical attenuation (loss) factors inherent to radio wave signal propagation; free-space attenuation, diffraction-, reflection- and scattering losses

Propagation models are separated and categorized by large-scale and small-scale models. Large-scale models predict the mean of the signal strength at the receiver, at an indicated distance from the transmitter. These estimates are advantageous in approximating the coverage area. Small-scale models, also referred to as fading models, are applied to describe the fast oscillations of the incoming signal over small distances and short time interims and takes into account variables such as time delay. To simplify the analysis on radio wave signal propagation, free-space losses are described in detail, whereas the losses from reflection, diffraction and scattering are briefly described.

3.4.1 Free-Space Losses

This section is adapted from Lambrechts and Sinha [5]. "Free-space propagation introduces losses in radio waves when travelling in LoS between the transmitter and the receiver. The free-space losses are proportional to the transmitted power and the distance that the signal travels towards the receiver. The Friis free-space equation defines the power received at the receiver (typically an isotropic antenna) and is given by

$$P_r(d) = \frac{P_{RAD}\lambda^2}{(4\pi)^2 d^2} \qquad (3.17)$$

where $P_r(d)$ is the power received by the receiving antenna (in watt) at distance d (in m), P_{RAD} is the equivalent isotropic radiated power from the transmitter and λ is the wavelength of the transmitted signal (in m). To account for additional effects such as the gain of the transmitting and receiving antennas as well as losses in the system and in the channel (medium) other than free-space, (3.15) can be rewritten such that

$$P_r(d) = \frac{P_t G_t G_r \lambda^2}{(4\pi)^2 d^2 L} \qquad (3.18)$$

where P_t is the absolute amount of power transmitted (also in watt), G_t is the gain of the transmitter and G_r is the gain of the receiver (dimensionless), and L is a random variable, the system loss factor ($L \geq 1$). A value of $L = 1$ would suggest that there are no losses in the system due to the overall contribution of transmission line attenuation, filter losses, losses in the medium or arbitrary antenna losses. It is important to note that from (3.17) and (3.18) the received power is reduced (attenuated) proportionally to the square of the distance d between the transmitter and the receiver. The signal losses due to free-space propagation of a LoS signal can be represented by a positive quantity as the difference between the received power and the transmitted power. If it is assumed that the transmitter and receiver antenna gains are unity, to eliminate the antenna properties in the loss equation, the signal losses of a LoS signal becomes

Table 3.3 Free-space path loss characteristics exponent n

Type of environment	Path loss exponent (n)
Free-space	2
Urban area	2.7–3.5
Shadowed urban area	3–5
In-building (LoS)	1.6–1.8
In-building (nLoS)	4–6
Factory (large indoor area)	2–3

$$PL(\lambda)[dB] = 20 \log_{10}\left(\frac{4\pi}{\lambda}\right) \tag{3.19}$$

which is only valid for values of d that exist in the far-field region and if $d \neq 0$. It can also be shown that the losses associated with the distance between the transmitter and receiver, is given by

$$PL(d)[dB] = 10n \log_{10}(d) \tag{3.20}$$

where n specifies the path-loss characteristics as a function of separation type between transmitter and receiver. Common values of n obtained from are given in Table 3.3.

In Table 3.3 it is shown that the path-loss exponent of free-space is equal to 2, and the path-loss exponent inside a factory or any large open indoor space is also approximated to 2 with maximum value approximated to $n = 3$. Urban areas and shadowed (high-density) urban areas have a path-loss exponent from $n = 2.7$ to a maximum value of $n = 5$. In-building LoS conditions present the lowest exponent values between 1.6 and 1.8 with in-building non-LoS (nLoS) conditions path-loss exponents approximated between 4 and 6. The effect of n is noticeable from Fig. 3.5, where the mean path loss of an 800 MHz communication signal over a distance of 5 km is simulated, for $n = 1, 2$, and 3."

As shown in Fig. 3.5, for an 800 MHz signal with a transmitter-receiver separation of between 0 m and 5 km, the signal attenuation varies significantly for different values of n. As n increases, the amount of attenuation experienced also increases, following the same type of attenuation, but shifted considerable on the signal attenuation axis in Fig. 3.5. There is also a noticeable difference between the increase of attenuation when n changes from 1 to 2 and when n changes from 2 to 3, showing a non-linear relationship as n increases. It is therefore recommended to keep n as low as possible, with free space ($n = 2$) being the lower limit (if not transmitted through a vacuum).

In terms of the wavelength/frequency of commonly-used commercial devices used to access the internet, several communication protocols and their corresponding operating frequencies are presented in Table 3.4. These frequencies can be used as reference to determine the expected signal attenuation of common communication

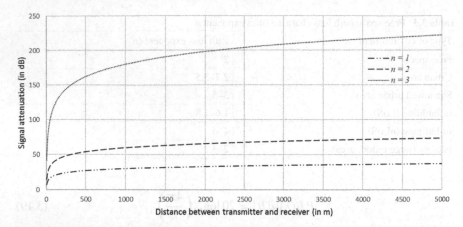

Fig. 3.5 Signal attenuation as a function of n for an 800 MHz signal at an arbitrary distance of 5 m between a transmitter and a receiver

Table 3.4 Approximate frequency bands for commonly used communication protocols in end-user devices used to access the internet—typical frequencies for delivering last mile wireless access (excluding mm-wave frequencies for Li-Fi and mm-wave backhaul technologies such as 5G)

Communication protocol	Frequency range
Wi-Fi	2.4 GHz and/or 5 GHz
GSM	Up to 3.5 GHz (3G)/up to 8 GHz (4G)
Bluetooth	2.4 GHz
NFC	13.56 MHz
ANT + UWB	2.4 GHz
WiMax	2–11 GHz
6LoWPAN	900 MHz/2.4 GHz
Zigbee	2.4 GHz

protocols. The protocols provided does not, at this point, include high-frequencies such as mm-wave and Li-Fi frequencies, as these will be presented separately.

The frequency ranges presented in Table 3.4 ranges typically from the 800 to 900 MHz band, to the commonly used 2.4/5 GHz bands for Wi-Fi and Bluetooth, as well as higher frequency implementations up to 8 GHz for 4G and 11 GHz for WiMax. Figure 3.6 demonstrates the relationship between the frequency (or wavelength) of a radio signal and the losses encountered in free-space at an arbitrary distance of 5 m between the transmitter and the receiver. The frequency band used as reference in Fig. 3.6 represents the commonly used frequencies in wireless communication devices for commercial equipment, such as end-user devices capable of accessing last-mile internet access, presented in Table 3.4.

As shown in Fig. 3.6, the expected signal attenuation/loss between a transmitter and a receiver separated by 5 m, varies according to the operating frequency in a

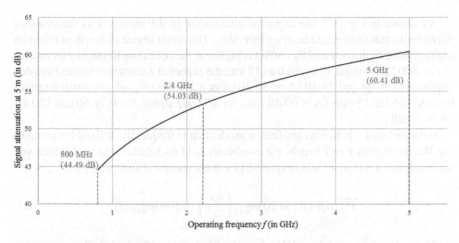

Fig. 3.6 Signal attenuation as a function of operating frequency (in GHz) for a frequency distribution of commonly implemented last-mile communication protocols at an arbitrary distance of 5 m between a transmitter and a receiver

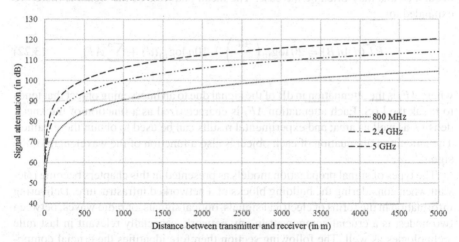

Fig. 3.7 Signal attenuation as a function of distance (in GHz) between a transmitter and a receiver at various commonly-used operating frequencies

logarithmic relationship. In the 800–900 MHz frequency band, typically used by older GSM networks, an attenuation of approximately 37.50 dB is expected over 5 m. This attenuation increases to 47.04 dB at 2.4 GHz and 53.42 dB at 5 GHz, still at a 5 m separation between transmitter and receiver. The expected signal attenuation can also be represented as a function of distance, a convenient method to determine the limitation in coverage area as a function of the link budget of the system. The attenuation as a function of distance, up to a practical separation distance of 5000 m (5 km) is presented in Fig. 3.7.

As shown in Fig. 3.7, the expected attenuation in dB increases as the distance between the transmitter and receiver increases. The offset attenuation is dependent on the operating frequency and this offset is higher as the operating frequency increases. For an 800 MHz signal, between 0 and 5 km, the expected attenuation ranges between approximately 44 and 74 dB ($\Delta \approx 30$ dB). For a 2.4 GHz signal, attenuation ranges between 54 and 114 dB ($\Delta \approx 60$ dB) and for a 5 GHz signal, between 60 and 120 dB ($\Delta \approx 60$ dB).

It follows that a path-loss prediction model that is frequently utilized for estimating the mean path loss (therefore a combination of its relationship to distance and wavelength) of a signal and is expressed by the subsequent formula,

$$\overline{PL}(d)[dB] = 20 \log_{10}\left(\frac{4\pi}{\lambda}\right) + 10n \log_{10}(d). \tag{3.21}$$

The attenuation model in (3.21), from Lambrechts and Sinha [5], "however, only accounts for LoS communication, and needs to be expanded for cases where nLoS occurs, as this may often be the case. The mean path loss of the signal is therefore expanded to

$$\overline{PL}(d)[dB] = 20 \log_{10}\left(\frac{4\pi}{\lambda}\right) + 10n \log_{10}(d) + \sum_i A F_i \tag{3.22}$$

where AF_i is the attenuation in dB of the signal due to an intervening object that tends to break the LoS. Each attenuation AF_i is characterized as a function of thickness, density, as well as area; and experimental results can be used to obtain these values. The amount of attenuation of each object is also a function of the wavelength of the signal."

The types of signal propagation models as presented in this chapters become relevant when considering the building blocks of a networked infrastructure. Delivering information in the forms of electrical signals, optical signals, or radio waves, between two nodes, is a crucial component in any network, especially relevant in last mile technologies as well. The following section therefore identifies the crucial components of any network system, with reference to the technologies discussed in the first part of this chapter. As these building blocks are identified and the alternative technology implementations acknowledged, it should become apparent that there are distinct advantages and disadvantages to each, in most cases directly related to the limitations of the propagation characteristics. However, before identifying these building blocks, the open systems interconnect (OSI) model [3] is briefly reviewed, as this model is often referenced to in any network infrastructure discussion.

3.5 The OSI Model

This section briefly reviews the conceptual framework model known as the OSI model and its abstraction layers. The goal of this section is not too dissect the model, but rather to provide an informative reference to the visually represented OSI model that will be used throughout this book and is used in this chapter to define the network building blocks. The OSI model can be referenced for most descriptions of a networking system, typically done when referring to last mile technologies. Each layer of the OSI model can be standardized and used for various applications and with emerging technologies such as Li-Fi and mm-wave backhaul technology, the OSI model serves as an ideal cross-reference to determine viable replacement technologies. An adaptation of the OSI model is given in Fig. 3.8.

The OSI model as shown in Fig. 3.8 is classified as an existing conceptual model that aims to characterize and also standardize the intercommunication among ubiquitous communication devices and hide the underlying internal infrastructure and technologies used to achieve the transfer of data. The OSI model is essentially partitioned into seven abstraction layers, sorted from the top (seventh) layer to the bottom (first) layer; which are the application-, presentation-, session-, transport-, network-, data link-, and the physical layer [12]. Each of these layers can be reviewed in extensive complexity and detail, but for this book, they are briefly described below.

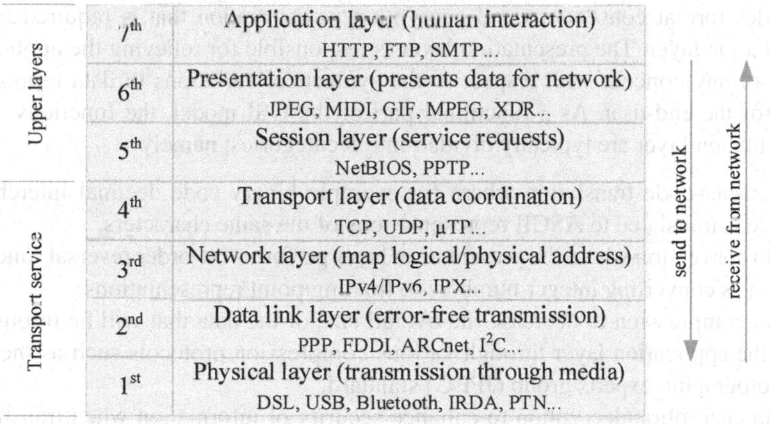

Upper layers	7th	Application layer (human interaction) HTTP, FTP, SMTP...	send to network
	6th	Presentation layer (presents data for network) JPEG, MIDI, GIF, MPEG, XDR...	
	5th	Session layer (service requests) NetBIOS, PPTP...	
Transport service	4th	Transport layer (data coordination) TCP, UDP, μTP...	receive from network
	3rd	Network layer (map logical/physical address) IPv4/IPv6, IPX...	
	2nd	Data link layer (error-free transmission) PPP, FDDI, ARCnet, I^2C...	
	1st	Physical layer (transmission through media) DSL, USB, Bluetooth, IRDA, PTN...	

Fig. 3.8 Visual representation of the conceptual framework known as the OSI model and its seven abstraction layers

3.5.1 Application Layer (7th)

The topmost, 7th, layer in the OSI model is the application layer and is what is presented to the end-user. The application layer is therefore a practical, user-friendly, and evolving layer that allows the end-user to interact with essentially the physical layer. Controls, inputs, and other human-interface devices connected to the application layer must be ergonomically designed and easy to interact with by human. In a network environment, the application layer supplies the network capabilities and services to the end-user application. These network services are usually set protocols that handle a user's data. Examples of the application layer include web browsers, instant messaging clients, mail clients or text editors. Protocols that are typically employed in the application layer include hypertext transfer protocol (HTTP), file transfer protocol (FTP), and simple mail transfer protocol (SMTP).

3.5.2 Presentation Layer (6th)

The presentation layer is the defined part that operates free of data representation in the topmost layer, and *"presents data for the application or network"* [12]. It is therefore used to both convert and prepare data in the application layout to the network design format, or vice versa. It handles syntax processing of message data which includes format conversions and encryption or decryption that is required by the application layer. The presentation layer is responsible for relieving the application layer of any concern with respect to the syntactical variations in data representation for the end-user. As a functional part of the OSI model, the functions of the presentation layer are typically divided into 5 categories; namely

- character-code translation where for example binary code decimal interchange code is translated to ASCII representations of the same characters,
- data conversion where the presentation layer performs bit-order reversal functions such as converting integer numbers to floating point representations,
- data compression to decrease the overall size of the data that will be throughput to the application layer through various compression protocols such as the joint photographic experts group (JPEG) standard,
- data encryption/decryption to enhance security of information when transmitted across a network, typically involving the secure sockets layer (SSL), and
- data translation to enable transparent interconnectivity between different types of computers, servers, and mainframes and fixing any irregularities in the transmission.

Protocols that are typically used within the presentation layer, which includes JPEG, are the musical instrument digital interface (MIDI), moving picture experts group (MPEG), transport layer security (TLS), graphics interchange format (GIF), and external data representation (XDR).

3.5.3 Session Layer (5th)

To create and ensure order when two or more devices need to *speak* to each other, the session layer is responsible for creating a unique session identification number. The session layer must reply to service requests from the presentation layer and also produce service requests to the transport layer. To generate a session, the session layer must define setup, coordination, and termination parameters for each session. This layer manages the order and flow of actions that start and ultimately terminate network connections and can handle several sorts of connections that can be established dynamically and communicate over discrete links. Session layer managing protocols are often referred to as application programming interfaces (APIs), where the network basic input/output system (NetBIOS) and point-to-point tunneling protocol (PPTP) are examples of session layer APIs. NETBIOS allows applications on individual and separate computers to communicate over a local area network (LAN).

3.5.4 Transport Layer (4th)

The transport layer confined in the center of the OSI model handles coordination of data to the suitable application process on the host computer. This includes determining the size of the data that must be transmitted, the agreed upon data rate, as well as the destination of the data. This layer is therefore primarily responsible for delivering the data across a network connection. Depending on the protocol used to deliver the data, additional capabilities could for example include error detection and recovery (typically through a checksum), flow control by implementing data buffers, and support for retransmission and automatic repeat requests. Commonly used transport layers include transmission control protocol (TCP), user datagram protocol (UDP), and micro transport protocol (μTP).

3.5.5 Network Layer (3rd)

If data reaches the network layer, it examines the sender and receiver addresses confined in separate frames and determines if the data has arrived at its target destination. If so, the network layer arranges the data into packets and directs it to the transport layer. If the destination parameters are inconsistent with the delivery address, the network layer changes the target address and shoves the data to the subordinate layers. The network layer is additionally accountable for mapping between logical (virtual) addresses and physical (media account control (MAC)) addresses which is required for correct data routing. The network layer essentially determines the most efficient path for the data to take while travelling between the source and the destination, and could route data globally through multiple paths, depending on what it deems the most efficient route. This layer is also capable of breaking up a message into smaller pieces, routing each piece through a different path, and reassembling it at its final destination. Store-and-forward is a typical method used in the network layer, where information is transmitted to an intermediate node and kept to be sent at a later stage, either to its final target or to an alternative transitional node. Commonly used protocols in the network layer include internet protocols (IPv4/IPv6) and internetwork packet exchange (IPX).

3.5.6   Data Link Layer (2nd)

The data link layer ensures that all packets of information in a physical transmission (from the physical layer) are transmitted free of errors. This layer additionally ensures that the correct physical protocol is consigned to the data. The data link layer has three primary functions, namely:

- handling transmission errors,
- regulating the flow of data, and
- ensuring a well-defined interface to the network layer.

   This layer therefore provides transfer of data in a node-to-node fashion and due to its relative complexity in the OSI model, it is delineated by dual sub-layers; the MAC layer and the logical link control layer (LLC). The MAC sub-layer runs services such as error as well as flow control, while the LLC sub-layer provides services for example multiple access control, LAN switching, data packet scheduling, store-and-forward, quality of service, and virtual LANs. Commonly used protocols defined as data link layer protocols include Ethernet, IEEE 802.11 wireless LAN, fiber distributed data interface (FDDI), asynchronous transfer mode (ATM), point-to-point protocol (PPP), attached resource computer network (ARCnet), I^2C, and many forms of serial communication.

3.5.7   Physical Layer (1st)

Finally, the physical layer of the OSI model is used for communicating and delivering the individual bits of data across media between the source and the receiver. Data are transferred using the form of signaling (wired or wireless) that is supported by the physical medium, such as electrical voltages, radio frequencies, pulses of infrared light, or pulses of regular light through optical fibers. Physical layer components therefore include

- the network interface controller (NIC)—as well as repeaters, Ethernet hubs, modems, and fiber media converters,
- connectors and interfaces, and

- cables such as shielded/unshielded twisted pair or coaxial cables.

The major functions performed by the physical layer include bit-by-bit delivery, modulation, bit synchronization, flow control, circuit switching, multiplexing, pulse shaping, and channel coding. Examples of the physical layer include digital subscriber line (DSL), Bluetooth physical layer, optical transport network (OTN), universal serial bus (USB) physical layer, infrared data association (IRDA), and Ethernet physical layers such as 10BASE-T.

As mentioned at the beginning of this section, the overview of the OSI model will be used as a cross-reference throughout this book and a low-level of understanding of the abstraction layers of this model is beneficial when discussing last mile solutions—especially for new generation technologies where certain layers are yet to be defined or standardized. Modern communication technologies demand high-bandwidth, efficient, and commercialized (standardized) technologies and protocols to serve multiple users with network and/or internet connectivity. Designing and maintaining an infrastructure that delivers last mile access should therefore consider these factors and provide solutions in a wide range of environments; particularly in developing countries and in rural areas, as is the focus of this book.

The following section reviews the primary network building blocks that are required to establish a network and serve users with information. These building blocks include the client computer or end-user device, the server computer, network interfaces, cables and wireless networks (physical media), switches, and networking software. Each of these building blocks are discussed, with reference to the OSI model and technologies that dominate modern internet access. The goal of this section is not only to identify the building blocks and the dominating technologies, but also to provide the reader with an overview of alternatives when considering novel implementations and infrastructures when considering last mile access.

3.6 Networking Building Blocks

The first network building block that has a significant impact on the type of last mile strategy and infrastructure, is the end-user device. Referencing the OSI model, the end-user device is part of the application layer, and should be able to decode information from a variety of sources (network protocols) and present it in an intuitive manner to the user. Typical end-user devices in modern times are discussed in the following paragraph.

3.6.1 Client Computer/End-User Device

The end-user is the person that a software application or an integrated hardware device is designed for. The consumer, or end-user, expects a service or application to be user friendly, practical, and provide capabilities that enrich their life through providing information or deliver a recreational service. The back-end development of the application should not be important for the end-user to know about or understand and this is originated, managed, and marketed by the developers, programmers and administrators. The end-user device is the device that displays the content in an easy-to-understand and interpretable way, without the end-user requiring a technical understanding of its workings.

Consumers and users of the internet have access to a large number of end-user devices, historically limited to the personal computer. A high level of competition among manufacturers have resulted in these many different types of devices and operating systems running in the back-end of these devices. The modern market structures of smart devices are relatively similar to that of the previous generation of personal computers. Computer operating systems (OS's) such as Microsoft's Windows, Apple's Mac OS and Unix-based operating systems have been the primary alternatives for users for several years. Modern smart devices such as smartphones and tablets primarily runs on Google's Android and Apple's iOS. Microsoft's mobile Windows OS, Blackberry OS and Nokia's Symbian OS have also had a fair share of the mobile OS market segment, but have virtually disappeared by the time of writing. For the general user, functionality and user experience is seemingly similar for both these operating systems; the differences mainly in the approach each company takes in managing and distributing each OS. Google's Android adopts an open-source strategy, making its OS freely available to manufacturers to change not only its look and feel, but also add or remove functionality they deem necessary to achieve their vision. Android has been described as fragmented as a result of this strategy, where updated versions of the OS is dependent on various factors such as the compatibility of the hardware, lead times for the manufacturers to update its current generation of devices and modifications made by service providers, further prolonging or preventing future updates. Apple's approach is more of an integrated approach. It aims to provide a high level of fluidity to its consumers, guaranteeing consistent upgrades in services, security, performance and protection of personal data. Apple controls access to its devices from third-party providers and have a strict editorial policy, ensuring high quality of applications and services being deployed on their OS. Although preference of which OS to use is based on the user, both modern operating systems allow for a different and future-proof means of accessing the internet through mobile devices and providing services to users of virtually infinite varieties.

End-user devices that grant the user access to information through the internet is dominated by smartphones and tablets, but there are various other types of devices that offer similar functionality. These include

- internet boxes that ads internet-access to television sets that do not have the functionality built-in, essentially converting it to a smart device,
- television set top-boxes that accesses certain internet-services and displays on a television monitor,
- personal computers and laptops,

- video game consoles,
- smart televisions, and more recently,
- connected wearables such as watches and fitness trackers.

From this list it should be noticeable that for manufactures, service providers and companies invested in last mile access, the types of services and the way that information is displayed to users have changed vastly in the past decade. The personal computer is not the primary target application and the services and applications should be altered based on how it will be presented to the user. Mobility has become the highest priority when designing and implementing last mile access in most circumstances, especially in developing areas where devices such as personal computers, video game consoles and smart televisions are less prevalent and considered a luxury item. In these areas, smartphones, and typically lower-tier and lower-cost versions thereof are abundant (with inherent limitations that should be considered). Furthermore, OS distribution in developing and developed countries differ significantly. Developing countries typically have more users using the Android infrastructures since there are less expensive alternatives available compared to Apple devices that are more premium. According to StatCounter [9], in April 2018 in South Africa alone, a country considered developed in the African context, 79.71% of smartphone and tablet users have Android devices and only 10.89% of users access online content using iOS devices. Other operating systems such as Windows, Blackberry OS, Tizen (from Samsung) and unknown devices make up the remaining 9.4%. In Europe, 71.46% of users are accessing online content through smartphones and tablets using Android, 26.55% through iOS, and the remaining 2% are from other operating system. Providers must take this into account when planning how to distribute information for large numbers of people in areas and environments that also have limited infrastructure for current- and future technology development.

3.6.2 Server Computer

Another crucial component in a network topology is the server computer, also referred to as the central unit. These components primarily provide the users with shared resources such as storage, internet access and services such as printing and off-site computational resources. These computers have specialized operating systems that manage user requests as well as specific software and services to provide networking services to the end-users. In Fig. 3.9, a schematic overview of the responsibilities within a network infrastructure is depicted.

As seen in Fig. 3.9, the server computer acts as a distributor of the internet to individual nodes (the end-user). The server is supplied with its own database, required to manage the connected users, and an administrator is granted access to the server computer to perform these management tasks. There are also various types of servers, which include

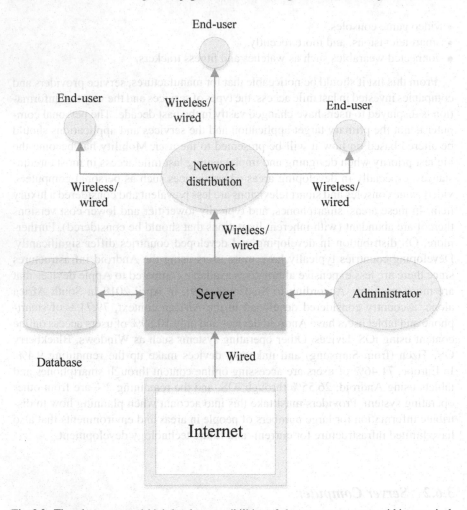

Fig. 3.9 The placement and high-level responsibilities of the server computer within a typical network infrastructure, adapted from Anderson et al. [1]

- web servers used to deliver static content to an internet browser, typically using protocols such as HTTP,
- application servers typically used to connect the end-user with a database server,
- real-time communication (RTC) servers that enables the instantaneous exchange of information through messaging services,
- FTP servers enabling secure transfer of files such as documents or media between clients,
- collaboration servers that allows users to collaborate and work together through for example the internet,

- list servers managing and delivering mailing lists, allow interactive discussions among users, or one-way communication to deliver newsletters, announcements or advertising,
- Telnet servers that enable users to remotely connect to a computer/server,
- open-source servers that give administrators and users open-source access to create application-specific services and tasks, and
- virtual servers which emulates the hardware and software requirements within a contained (sandboxed) environment.

The connection between the internet and the server is a critical path and the focus of this book, therefore access protocols, standards and novel technologies should be accommodated in this path. A central server, although typically the de facto standard to enable network distribution to end-users, also have various advantages in managing and maintaining the last mile access to the users. A high-performance architecture, capable of running various tasks in parallel, as a standalone system, is advantageous for reasons such as (adapted from Anderson et al. [1]):

- these devices can take advantage of specifically-designed infrastructures to accommodate the facility and immediate environment. Aspects such as permanent power sources and backup power through renewable sources provides the server with uninterrupted power,
- ownership of the server computer is typically by a single entity,
- the server computer acts as a single contact point for the network infrastructure and connects the end-users to the respective service providers without the user having to choose the best route at a given time,
- both physical and network security can be implemented, managed and maintained by skilled workers,
- restrictions of unauthorized services can be managed by the server,
- housekeeping and maintenance of the server can be performed by installing sensors and actuators that monitor the status of the server at any given time, allowing for preventative maintenance and allowing uninterrupted network and internet access to the end-users,
- standardization of server computer software and resources is a well-established sector and network service delivery to the end-user can be performed by implementing common standards such as Wi-Fi, and
- back-end network optimization and improvements to last mile access can be performed without affecting the communication requirements between the end-user and the network infrastructure.

The listed advantages are not exclusive to the proposed last mile topologies that will be presented in this book; however, it is important to take note of the importance of a central server computer in a network topology. Since much of the proposed technologies in this book, such as Li-Fi and mm-wave backhaul networks, are considered future generation technologies, it remains critical that the service delivery to end-users remain uninterrupted and that the capabilities of the server are chosen to provide seamless integration of these newer technologies. There are off course also certain limitations and disadvantages of using central server computers, however it is typically encouraged to mitigate these as opposed to not using a central server in a network topology. These include

- the capital investment of the server computer along with its installation, supporting components such as cables and human interface devices is typically high,
- managing of large networks and a large number of services is complex and require skilled workers that are available any time the server is experiencing downtime,
- a central server is essentially a single point of failure and any downtime affects multiple end-users,
- file servers can become corrupt or physically damaged and replacing or repairing it is not only costly, but could lead to high volumes of data loss,
- viruses and malware attacks loaded on the server can easily spread to end-users, and
- hacking is typically focused on server computers as opposed to end-user devices due to the large volumes of data and user-sensitive information that it may contain.

As mentioned, the disadvantages of using a server computer typically do not outweigh the advantages and with proper planning, regular maintenance and skilled workers assigned to manage and maintain the equipment, the use of a server computer as a building block in a network infrastructure is crucial.

3.6.3 Network Interface/Switches

As a member of the networking building blocks, the network interface (traditionally the NIC when the circuitry was not installed as a standard feature on most computers) refers to the combination of hardware and software that enable communication between devices. The network interface can either be a physical device, or refer to a virtual representation that enable communication on specific standards or protocols. Within the OSI model, the network interface is a physical layer and data link layer device since it establishes a network node and must have a unique physical address (MAC) [12].

A network interface is used to connect devices within various infrastructures, among the most common being the LAN and wide area network (WAN). A LAN is a portion of the network that is contained by geographical boundaries and typically used in home or office networks. Importantly, a LAN is not publicly accessible to users outside of the network, for example through the internet. Conversely, a WAN describes large, dispersed networks and are often interconnected through the internet, accessible to users from virtually anywhere depending on the administrative requirements.

Any network interface implements protocols, a set of rules that must be followed to establish a connection and allow the transfer of information. The specific protocol used at any given time also depends on the network layer, since multiple technologies and protocols are needed to provide the necessary resources to communicate over an interface. Examples of low-level protocols include TCP and IPv4/IPv6 and these enable higher-level protocols (such as the application protocol) for example HTTP to transfer web content to users. The most widely used network connection for personal computers is an Ethernet connection and most modern computers, servers, and mainframes have the NIC built into the motherboard of the system.

3.6.4 Physical Media (Cables and Wireless Networks)

The physical media that connects a transmitter and a receiver, therefore the source generating the data/information and the destination, are either wired or wireless, depending on the environmental/geographical limitations, end-user application, and the overall infrastructure that could affect the cost of implementing either of the two technologies. In developing country last mile access, the option of using cables or wireless to deliver internet access to central points where the users can access the network, depends on many factors, and it is the goal of this book to explore these alternatives. At first, it would seem that wireless access is ideal, since it eliminates the need to physically install cables within the environment. Wired access can be cost effective if implemented correctly, however, in many developing countries, issues such as cable theft, vandalism, and lack of maintenance hinder the development of high-bandwidth networks for users in rural areas. This section provides a brief review of the available wired and wireless technologies that can be applied for internet distribution and last mile access.

In terms of wired alternatives, the type of cables that can be used depends on the technology used to distribute network and internet access from the service providers. The typical types of cables that are used to connect nodes on a network are

- unshielded twisted pair (UTP) cabling,
- shielded twisted pair (STP) cabling,
- coaxial cables,
- power lines, and
- fiber optic cables.

In terms of twisted pair cables, both shielded and unshielded, these are the leading cable standard for Ethernet connections, containing up to eight wires wound together in pairs to minimize electromagnetic interference (EMI). Most modern Ethernet cables utilize UTP due to the cost savings and the acceptable performance it offers over short and medium distance, where STP are usually implemented in higher bandwidth distributions such as a FDDI. STP provides additional shielding (screening) to further prevent EMI by providing an electrically conductive barrier to attenuate EM waves that are generated externally from the shield, typically from nearby sources. This shield thereby provides a conduction path for the externally generated signals to return the currents to the source through a ground connection. Figure 3.10 provides a visual representation of the external EMI as it radiates towards a shielded cable, demonstrating how the radiation would enter and exit the cable.

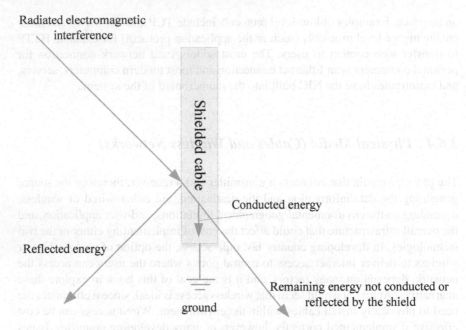

Fig. 3.10 The distribution of incident EM radiation on a shielded cable, showing the components that are reflected, conducted, and remaining in the system

As shown in Fig. 3.10, there are typically three components associated with the EMI as radiation from an external source is incident on a cable. The reflected energy is a function of the quality of the shield and should be maximized as this is the component of EMI that is not able to enter the cable due to the shielding. There will inherently be a component that enters the cable, and is conducted by the shield, which is ideally grounded, and should not interfere with the primary signal within the cable. The remaining energy that is not conducted or reflected by the shield could have adverse effects on the primary (wanted) signal generated inside the cable, and should ideally be minimized. There are three commonly-used shielding strategies, which includes

- Individual shielding where an aluminum foil screening is placed over each individual cable therefore eliminating crosstalk and EMI between each individual cable. This implementation is the most expensive solution but provides the highest level of EMI attenuation.
- Overall shielding where the screening material (either aluminum foil or metallic braided shields) is placed over pairs of cables, since the windings pairs are already designed to limit EMI.
- Individual and overall shielding which covers the twisted pairs and provides an outer shielding over all cables, reducing the EMI from external sources as well as among twisted pairs inside the cable.

The primary difference in the shielding materials, either aluminum foil or metallic braided shielding, lies in its ability to provide mechanical strength, flexibility effectiveness at certain frequencies. The type also influences the cost of the cable, where braided shields are typically used as military specification and are more expensive compared to the lightweight foil alternatives, most used in commercial applications. In terms of last mile access, the type of shielding is important to limit EMI in medium to long distance cables, however, it should also be noted that the amount of external sources that generated EMI around the area are typically less of an issue in developing countries and rural areas. The type of cable should therefore be chosen by keeping in mind the immediate environment rather than choosing the most expensive solution that should provide the most EMI shielding. The difference between an individual shielding and overall shielding might be negligible when consider the distance and the neighboring environment, but it would drive the cost up significantly. Furthermore, electrical signals are also subject to electrical losses, as discussed in Sect. 3.2.

3.6.4.1 Fiber Optic Cables

Fiber optic cables have several advantages when compared to traditional copper-based cables. In modern connectivity infrastructures, fiber has gained a lot of traction as the preferred solution due to these advantages, although in most circumstances still more expensive that traditional solutions, the benefits of using fiber are not only beneficial towards current solutions, but also for future-proofing bandwidth capabilities. The dominant factors of fiber that has led to its success and rapid uptake include

- the high-bandwidth capabilities,
- its immunity to EMI,
- no corrosion over time,
- resistance to intersection (eavesdropping),
- its ability to be used in conjunction with older technologies, and
- its physical attributes—being a lightweight solution.

A fiber optic cable comprises a single, tight-shielded fiber bounded by additional materials to enhance the strength of the cable, since the individual fibers are extremely fragile and inherent to bending which causes signal loss. In Fig. 3.11, the typical construction of a fiber optic cable is presented.

The core of the fiber, as shown in Fig. 3.11, is a transparent glass component allowing optical signals to pass from the transmitting side of the cable towards a receiver at the other end. The cladding is a glass sheath surrounding the core and reflects light back into the core. Light therefore bounces off from the cladding as it propagates through the cable in various modes, as discussed in Sect. 3.3. Additional buffer layers, as shown in Fig. 3.11, further enhance the strength and bending capabilities of the cable. There are five types of fiber optic cables to consider when implementing as a solution for connectivity (last mile or any other application). These types include

- plastic cables for inexpensive solutions over short distances,
- plastic-coated silica cables offering somewhat better performance than plastic cables,
- single-index single-mode cables for long-distance and high-bandwidth implementations,
- step-index multi-mode cables for cost-effect medium distance applications such as LAN environments, and
- graded-index multi-mode cables to increase the covering distance of step-index solutions, while still allowing for multiple light-source communications and bandwidth enhancements.

The core and cladding specifications of fiber cables (also described in Sect. 3.3), are also considered based on the specific application. For example, a FDDI solution has a minimum recommended core to cladding ratio of 62.5/125 μm and requires a multi-mode fiber cable. Furthermore, the following guidelines in terms of core and cladding sizes are presented;

- core stipulations for step-index as well as graded-index multi-mode wiring is defined for a series from 50 μm to 1 mm,
- the width of the cladding for step mode wiring is defined from 125 μm to 1.05 mm, and
- the core/cladding width for single-mode step cables is 4–10 μm/75–125 μm.

Optic fiber cables also implement a color-coding scheme to distinguish among cable types and identify individual fibers or groups of fibers within a cable. There are

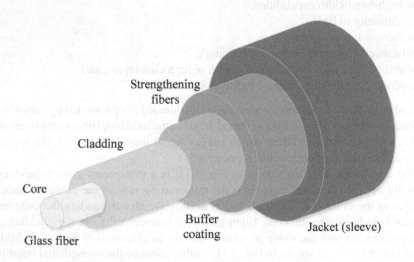

Fig. 3.11 The simplified physical construction of an optic fiber cable

several standards that describe the color-coding identification systems, where these standards include

- the TIA/EIA-598 (Bellcore) standard,
- S12,
- Standard Type E, and
- FIN2012.

All schemes are described by realizing 12 different colors to classify fibers which are clustered together in a collective package such as for example ribbon or tube (or yarn in certain circumstances). It is important to identify the standard used when installing and maintaining fiber infrastructures and keep track of the color-coding system. Skilled workers are typically sent on training courses to understand and use these systems. There is also various online source available that describe each of these color-coding standards and a description of each is not presented in this book.

Ethernet technologies for transmitting Ethernet frames at different data rates, light waves, and coding techniques are defined by, for example, the IEEE 802.3 set of standards. In terms of optic fiber cables and Ethernet solutions using fiber, Table 3.5 summarizes the most commonly-used solutions for fiber infrastructures, adapted from Tan [11].

From Table 3.5, 10BASE-FL refers to the 10 Mbit/s Ethernet standard using fiber optic cables, where FL denotes fiber link. 10BASE-FL is currently rarely used and has almost entirely been replaced by Fast Ethernet, Gigabit Ethernet, and 100 Gigabit Ethernet standards. 100BASE-FX is such a version of Fast Ethernet over fiber, using a 1300 nm near-infrared (NIR) light-source and allows for 100 Mbit/s transmission at a maximum distance of 2 km (for full-duplex over multi-mode fiber). 100BASE-SX and 100BASE-LX are variants of this 100 Mbit/s standard, using light sources at 850, 1310, or 1550 nm and each light source used for either upstream or downstream transmissions. The 1000BASE SX, -LX, and -LH variants enable 1 Gbit/s transmission (Gigabit Ethernet) at various distances, also implementing light sources from 850 up to 1300, 1310 and 1550 nm and are popular solutions for

Table 3.5 Various fiber optics standards as defined by the Institute of Electrical and Electronic Engineers (IEEE 802.3 standards)

Standard	Data rate (Mbps)	Cable type	Maximum distance (km)
10Base-FL	10	MM, 850 nm, a or b	2
100Base-FX	100	MM, 1300 nm, a or b	2
100Base-SX	100	MM, 850 nm, a or b	0.3
100Base-LX	100	SM, 1310 nm, 1550 nm, c	100
1000Base-SX	1000	MM, 850 nm, a or b	0.22–0.55
1000Base-LX	1000	MM, 1300 nm, a or b SM, 1310, c	0.55–2
1000Base-LH	1000	SM, 1550 nm, c	70

Legend $a = 50/125$ μm, $b = 62.5/125$ μm, $c = 9/125$ μm

LAN solutions in large office blocks, or long-distance transmission in the case of the 1000BASE-LH standard.

3.6.4.2 Wireless Protocols

Wireless protocols for end-users enable devices to access internet services and other connectivity-related services without the need for a physical cable. Modern devices are capable of using multiple wireless standards to enable this functionality, and last mile internet connectivity for the end-user can be achieved in various ways. The most common protocols that are found on many end-user devices include

- Wi-Fi,
- GSM access (GPRS, 2G, 3G, 4G/LTE and the modern, albeit less widespread, 5G networks),
- Bluetooth,
- near-field connections (NFC),
- ANT + ultra-wideband (UWB), and for certain applications also
- WiMax,
- 6LoWPAN, and
- ZigBee.

Each of these wireless communication standards listed above have distinct sets of advantages and disadvantages, typically with respect to transfer speed and range. However, in many practical applications, the popularity of the wireless standard becomes one of the most important factors to consider. Especially in developing markets and rural markets, the popularity of the wireless standard must be considered as a high priority. Wi-Fi, GSM, and Bluetooth are possible among the most influential candidates for delivering last mile access as most end-user devices are capable of communicating on these standards; NFC and ANT + UWB also have qualities that make them ideal for specific application, typically short-range high-speed data transfer. Advances in Wi-Fi technology, especially considering protocols such as the 802.11 a/g/n variants are capable of maximum bandwidths per channel of 54–270 Mbps, which should in theory be sufficient for serving multiple users at once. Efficient network design and signal distribution can ensure seamless and high-quality internet access by using Wi-Fi as the communication standard between the end-user and the service provider. In consequent chapters of this book, the primary differences (technical and practical) with reference to feasibility for last mile access are discussed in further detail.

In developing countries, issues such as cable theft and long distances among nodes require a reevaluation of using cables to deliver last mile access, and as will become apparent in this book, wireless technologies (radio waves, mm-wave radio waves or light fidelity) are among the recommended technologies to use as opposed to traditional cable distribution. Even high-bandwidth fiber networks can be subject to various limitations and difficulties to expand networks in these environments. Wireless technologies also offer carriers, service providers, and public/private networks the alternative of high-capacity connectivity through permanent point-to-point or point-to-multipoint communication or from a single or multiple locations [4]. In urban areas, service providers typically employ last mile solutions to deliver internet access from a communications provider to a consumer, using either cable of wireless technologies. As an example, a high-site (tower) will use wireless technology for last mile access and fiber networks to distribute data to and from the core network (back-hauling). A large number of network links in urban areas are backhauled from the fixed core network through fiber cables as opposed to wireless technologies. Wireless distribution is essentially more cost-effective to distribute network access, however, in most circumstances, (fiber) cables still offer superior transfer rates, especially between the core network and the service provider. In this book, it will be proposed that new generation wireless technologies are in fact capable of serving users with high-bandwidth internet access, and with reference to the theme of rural/developing markets, could possibly be a better solution than traditional cable networks.

3.7 Conclusion

This chapter identifies important considerations for planning and implementing last mile solutions, specifically with respect to signal propagation through wired and wireless media as well as the common building blocks of a network infrastructure.

Any signal, wired or wireless, is subject to losses as it travels between a transmitting source and a receiving node. These losses are dependent on various factors, primarily the distance between the transmitter and the receiver, the type of medium it is passing through, and effects that hinder the flow of information (as electrons, radio waves, or photons). There are therefore limitations with respect to the maximum distance between a transmitter and a receiver as well as the maximum bandwidth/throughput that a technology can offer, and these limitations influence the planning and implementation phases of a network. In last mile solutions implemented in areas with geographical limitations or where cost-effective but reliable solutions are required, these propagation factors must be carefully considered and accounted for. Signal losses and low signal integrity are factors that are not always noticeable to the end-user but can have a detrimental effect on the delivery of information to the consumer. Low speeds, network congestion, and unreliable connections are typical consequences of subpar planning and deployments. The first three sections of this chapter is therefore dedicated to reviewing expected losses in the three main types of transmission media, electrical signals, optical signals, and radio waves. It is recommended that the reader familiarize themselves with these concepts, as this forms the baseline for understanding how signals propagate and what the limitations are. Future chapters of this book focuses on new generation technologies such as Li-Fi and mm-wave backhaul technology and the limitation in these technologies are based on the same principles. The provided theoretical background in this chapter is sufficient to generate basic simulation models to predict the maximum distance (and the link budget) of typical infrastructures—which should give a relative idea of which technologies (electrical, optical, or radio waves) can be implemented in a proposed scenario. Since each infrastructure is different, especially in rural communities where many geographical factors lead to specialized and unique solutions, performing a preliminary simulation of a proposed infrastructure ultimately could save significant time and money on such a project. Furthermore, the future generation technologies discussed in this book will assume that the reader is familiar with the basic principles of signal propagation and use these to review these technologies in further detail.

The remaining sections of this chapter identifies the primary networking building blocks with reference to the OSI model. These building blocks typically comprise of hardware and software components that implement basic propagation principles, and upgrading or replacing these modules require knowledge of the factors that incur any limitations. These building blocks include the client computer or end-user device, the server computer, the network interface or switches, and physical media. Components such as the end-user device rely on receiver sensitivity to receive wireless signals; transmitted with a specific link budget across a variable distance—relating back to propagation principles and limitations. Server computers and network interfaces must be able to pass data and information on to the end-user through physical media, again relying on efficient and effective transfer of electrical, optical, or radio wave signal. The building blocks of any network, from an engineering perspective, are therefore as important as the fundamental principles of data transmission and the reader should be comfortable to identify these building blocks and the tasks performed by each.

References

1. Anderson R, Blantz E, Lubinski D, O'Rourke E, Summer M, Yousoufian K (2010) Smart connect: last mile data connectivity for rural health facilities. In: Proceedings of the 4th ACM workshop on networked systems for developing regions, 15 June, vol 5
2. Brillouin L (1922) Diffusion de la Lumière et des Rayonnes X par un Corps Transparent Homogéne; Influence del´Agitation Thermique. Ann Phys 17:88
3. Day JD, Zimmermann H (1983) The OSI reference model. Proc IEEE 71(12):1334–1340
4. Kotze C (2012) Last-mile wireless solution for restrictive rural areas. Retrieved 3 June 2018 from http://www.engineeringnews.co.za
5. Lambrechts JW, Sinha S (2017) SiGe-based re-engineering of electronic warfare subsystems. Springer International Publishing. ISBN 978-3-319-47402-1
6. Liebe HJ, Manabe T, Hufford GA (1989) Millimeter-wave attenuation and delay rates due to fog/cloud conditions. IEEE Trans Antennas Propag 37(12):1612–1617
7. Perez-Vega C, Garcia JL, Higuera JML (1997) A simple and efficient model for indoor path-loss prediction. Meas Sci Technol 8:1166–1173
8. Sklar B (1997) Rayleigh fading channels in mobile digital communication systems part I: characterization. IEEE Commun Mag 35(7):90–100
9. StatCounter (2018) Mobile operating system market share in South Africa—April 2018. Retrieved 7 May 2018 from http://gs.statcounter.com
10. Tamura Y, Sakuma H, Morita K, Suzuki M, Yamamoto Y, Shimada K, Honma Y, Sohma K, Fujii T, Hasegawa T (2018) The first 0.14-dB/km loss optical fiber and its impact on submarine transmission. J Lightw Technol 36(1):44–49
11. Tan TC (2000) Gigabit ethernet and structured cabling. Electron Commun Eng J 12(4):156–166
12. Yemini Y (2000) The OSI network management model. IEEE Commun Surv Tutor 3(1):20–29

Chapter 4
A Theoretical Analysis of Li-Fi: A Last Mile Solution

Abstract A last mile technology that is identified to potentially relieve the digital divide in emerging markets, and especially relevant in rural communities, is light fidelity (Li-Fi). Li-Fi can take advantage of current infrastructures (lighting) to distribute the internet to local communities in these areas. Li-Fi last mile solutions do not require large investments in implementing traditional infrastructures such as fibre or copper which are challenging, expensive or unpractical in these areas. Li-Fi uses light in the visible spectrum and propagates through free space—two important factors when considering an infrastructure that already exists in many rural communities. Information can be modulated onto these light signals (carriers) and be detected and demodulated with photodetectors. Since the (carrier) frequency of the light signal is very high (in the THz range), there are little bandwidth limitations when transmitting information using light, a distinct advantage of Li-Fi that offers numerous benefits. The principles of operation (including an introduction into Li-Fi channel modelling), benefits and limitations, as well as potential Li-Fi applications, are researched in this chapter and delivered to the reader as methodological principles of light-based communications.

4.1 Introduction

Light emitting diodes (LEDs) have become commonplace as a primary source for indoor lighting and are being implemented in various outdoor lighting applications with its brightness surpassing traditional fluorescent or incandescent sources. These sources of light have opened a new pathway for linking mobile devices to the Internet and present potential for wider bandwidth and faster response times when compared to traditional Wi-Fi. Visible light communication (VLC) or optical wireless communication (OWC) [2] and even applications using ultra-parallel VLC (using multiple colors of light to provide higher bandwidth linkages) are poised to supplement or in certain scenarios replace traditional radio-based communications. As with any emerging technology there are several challenges that limit the throughput and efficiency of a light-only communications system, and a VLC system such as light fidelity (Li-Fi) is no different. Li-Fi is the network solution for VLC that is

© Springer Nature Switzerland AG 2019   109
W. Lambrechts and S. Sinha, *Last Mile Internet Access for Emerging Economies*,
Lecture Notes in Networks and Systems 77,
https://doi.org/10.1007/978-3-030-20957-5_4

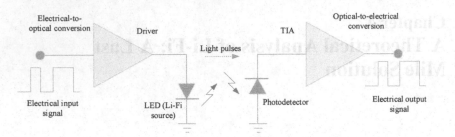

Fig. 4.1 Diagram of a communication network in the optical domain and the essential components required to convert from the electrical domain to the optical domain, and back again to electrical signals in a Li-Fi system

proposed to integrate with existing radio frequency (RF) access technologies. The primary components that determine the quality and speed of information transfer in a VLC system are the LED transmitter and the detector. Improvements in LED technology are rapidly changing the characteristics of these systems and the use of high-bandwidth photodetectors is ensuring that throughput increases to practical rates. The use of avalanche photodetectors, which produce a cascade of electrons from a single photon striking the photodiode, is further increasing bandwidth capabilities. In Fig. 4.1, a simplified representation of an optoelectronic communication system capable of converting electrical signals to optical signals—to be transmitted across a Li-Fi link—and back to electrical signals, is presented.

The VLC system in Fig. 4.1 consists of two primary sections, firstly, the section that converts digital information (electrical signals) to optical signals through a driver and LED light source. The information is then modulated and pulsed through free-space as light waves towards a photodetector that is sensitive to a specific wavelength of light, for Li-Fi, in the visible spectrum. The sensitive wavelength of the photodetector coincides closely with the wavelength at which the information is transmitted from the LED source. As the light waves are detected, small electrical signals occur in the photodetector that are amplified by a transimpedance amplifier (TIA) to produce a larger, usable, electrical signal and converted back to its digital form, where the end-user is then able to read the information sent through the channel.

In this chapter, specific focus is placed on the devices that are required to receive and manipulate the optical signals. The receiving devices (photodetector in conjunction with an amplifier) are importantly located on the end user's device and should be low power, small, mobile, and capable of eliminating noise and interference from nearby sources. The first device discussed in this chapter is the photodetector, a device that internally converts optical signals (photons) to electrical signals (currents) an electric field is present. The type of photodetector used in Li-Fi should have a high gain to enable the device to distinguish between information-containing light and ambient sources without being overexposed and unable to perform effective amplitude modulation of the incident light signals. The following section reviews how visible light is absorbed/transmitted through a medium. The typical channel of a Li-Fi solution is free space, and the following section highlights the propagation characteristics of light through such a medium.

4.2 Visible Light Absorption

As Li-Fi uses light from the visible spectrum, it is of importance to determine the absorption of visible light and to quantify the interaction of visible light with the medium through which it typically propagates. An ideal environment can be produced that allows light to travel between the transmitter and the receiver with limited attenuation through the channel. Visible spectroscopy studies the interaction of radiation (with wavelengths between approximately 380 and 750 nm) with a chemical species, therefore the medium through which it travels (typically free space). Since light moves in envelopes of energy, collective termed photons, each of these photons is associated with a specific energy associated to its wavelength. Absorbance is the amount of light that is absorbed by the medium/channel and is graphically represented in Fig. 4.2.

As shown in Fig. 4.2, the incident light travels through the sample (channel) and the transmitted light gives an indication of the transmittance, or its inverse, the absorbance. The ratio is defined by Horecker [7]

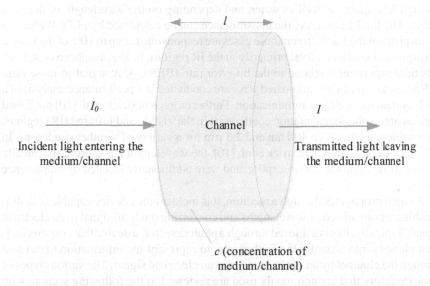

Incident light entering the
medium/channel

Channel

Transmitted light leaving
the medium/channel

c (concentration of
medium/channel)

Fig. 4.2 A graphical representation of absorbance where incident light enters a chemical sample, typically the channel through which light travels and the ratio of transmitted light quantifies the transmittance of the sample

$$\log_{10} \frac{I_0}{I} = \alpha cl \tag{4.1}$$

where I_0 is the intensity of the incident light, I is the intensity of the transmitted light, α is the specific absorption coefficient, c is the concentration of the absorbing material, and l is the length of the light path through the sample (channel). This relationship is only valid when monochromatic light is used and if the channel does not contain any absorbing impurities. From (4.1), the absorption coefficient is wavelength-dependent, and is determined by

$$\alpha = \frac{4\pi k}{\lambda} \tag{4.2}$$

where k is the extinction coefficient as a function of scattering, reflection, and aerosols in the atmosphere, and λ is the wavelength of the light. Absorbance can be measured using a spectrometer that comprises a source that emits light, concentrating lenses, a diffraction grating to fragment light into altered frequencies or wavelengths, and a photodetector that is capable of measuring the light passing through the sample. The detected light also undergoes a series of electrical amplification, typically using a TIA.

Since Li-Fi is typically used with free space being the channel or medium through which the information passes, the absorption of visible light due to chemical compounds in free space is of interest. Visible light typically has a high transmittance through free space, as well as water, and depending on the wavelength of the light radiated by the LED source, the transmittance can be calculated by (4.1). Within the atmosphere of the Earth, greenhouse gases are responsible for up to 70% of the known absorption of sunlight (albeit primarily in the IR region). In Shamsudheen et al. [19], practical experiments related to the bit-error rate (BER), peak signal-to-noise ratio (SNR), as well as the mean squared error are conducted in a performance analysis of a VLC system for optical communication. Furthermore, Knestrick et al. [10] published various atmospheric attenuation coefficients in the visible and infrared (IR) regions, for wavelengths between 400 nm and 2.3 μm for a variety of weather conditions. In this study presented by Knestrick et al. [10], the wavelength bands were specifically chosen to avoid molecular absorption and were additionally isolated by interference filters.

As light propagates through a medium, it is incident on a device capable of distinguishing among different wavelengths and converting the light signal to an electrical signal. Typically, this is achieved through a photodetector, a device that converts incident photons into a small electrical current to represent the information (data) sent through the channel by an optical carrier as an electrical signal. The various types of photodetectors that are commonly used are reviewed in the following section, with the preferred solution for a Li-Fi system highlighted.

4.3 Detecting Light Signals

To detect incident light, photodetectors have been used for a long time, with specific types of photodetectors having advantages and disadvantages in various applications. In this section, the primary differences among the most common (and least complex) *pn*-junction photodetector, the improved *p-i-n* photodetector, and the high-gain avalanche photodetector are discussed in lieu of the proposed application, being a Li-Fi receiver.

4.3.1 pn-Junction Photodetectors

Photodetectors are used to detect incident light on a receiver subsystem and are manufactured as various alternatives, where the *pn*-junction photodetector is the easiest to manufacture, albeit having several drawbacks due to its simplicity. Photodetectors absorb incident photons or excited elements and produce a flow of current that is relative to the incident energy or power on the device. Photodetectors can be used to detect minute magnitudes of light and its precision can be regulated from intensities below 1 pW/cm^2 to more than 100 mW/cm^2. One such device is the planar photodetector, which is essentially a photosensitive *pn*-junction diode, made by either diffusing a *p*-type impurity into an *n*-type substrate, or vice versa. The diffused zone outlines the active area where photons from light are received and the variation in electric field is transformed to energy in the electrical domain. A simplified representation of a *pn*-junction photodiode is presented in Fig. 4.3.

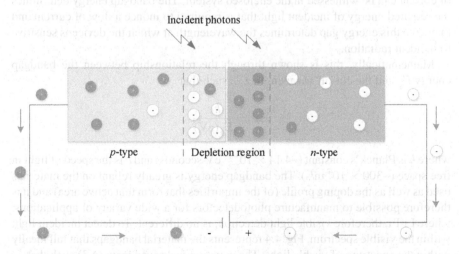

Fig. 4.3 Simplified representation of a *pn*-junction photodetector

To achieve an ohmic contact to the *pn*-junction photodetector presented in Fig. 4.3, additional impurity diffusions are performed at the top and bottom of the substrate (or the left and right with respect to Fig. 4.3). For a *p*-type active area, the ohmic contact is typically from an *n*-type impurity and for an *n*-type active area; the ohmic contact is from a *p*-type impurity. Contact pads are deposited at the positions where external contact (such as probing or wiring) is required. The active area is also passivated with an antireflection coating (ARC) to decrease the reflection of photons at particular wavelengths. The width (measured as the thickness in a side-view) and the type of material used for the antireflective coating govern the optical characteristics of the ARC and should be carefully applied to compliment the intended operation of the photodetector. The areas around the active area are physically protected by a layer of metal, which also reduces the probability of stray light entering the active area and generating unwanted spurious signals.

Apart from the physical construction of the *pn*-junction photodetector, the choice of materials used for the device is important in determining the wavelengths of light at which the device is sensitive. Since Li-Fi operates in the visible spectrum of light, photodetectors used in Li-Fi applications must be able to accept and convert incident light that contain information. The material bandgap is the opening in the middle of the valence band and the conduction band of the diode. At 0 K, also termed the absolute zero temperature, the valence band is entirely occupied and the conduction band is empty of any potential charge carriers. As energy is added to the system with an increase in temperature, electrons in the valence band become excited and are able to move to the conduction band. Not only through thermal energy can this be achieved, but also by external stimulation with for example an electric field (applied bias voltage) or by energy in photons that are higher than the bandgap energy. If an electron is promoted to the conduction band, it is free to conduct charge and a flow of current can be witnessed in the enclosed system. The bandgap energy determines the required energy of incident light that is required to induce a flow of current and therefore this energy gap determines the wavelength at which the device is sensitive to incident radiation.

Mathematically, this is shown through the relationship between the bandgap energy E_g and the cut-off wavelength λ_{co}, such that

$$\lambda_{co} = \frac{hc}{E_g(eV)} \tag{4.3}$$

where h is Planck's constant (~4.14×10^{-15} eV-seconds) and c is the speed of light in free space (~300×10^6 m/s). The bandgap energy is greatly reliant on the materials used as well as the doping profile (of the impurities that form that active area) and it is therefore possible to manufacture photodetectors for a wide variety of applications, where Li-Fi, therefore visible light detection, is no different. To detect incident light within the visible spectrum, Fig. 4.4 represents the material bandgaps that fall ideally within the spectrum of visible light. The equation is derived from (4.3) such that

$$E_g(\text{eV}) = \frac{hc}{\lambda_{co}}. \tag{4.4}$$

Importantly, the bandgap of the material only determines the peak response frequency and a material that falls outside this range can still be used to detect incident light.

As shown in Fig. 4.4, the ideal range of photodetector material energy bandgaps in eV to detect visible light lies between approximately 3.26 and 1.65 eV. As mentioned, the values of E_g presented are only the peak responsivity values and photodetectors that fall within these ranges can also be used in conjunction with optical filtering. A common responsivity and wavelength curve for Si photodetector is presented in Fig. 4.5.

As shown in Fig. 4.5, the peak responsivity of the (Si) semiconductor material is at a wavelength of approximately 950 nm. This means that the semiconductor material is capable of distinguishing between incident light *around* this wavelength, with its best performance at 950 nm. However, the responsivity (in units mA/W) around 950 nm is not zero, therefore this photodetector is also capable of recognizing incident light at other wavelengths. By using optical filters to attenuate wavelengths that are not of interest and/or mirrors to provide sufficient gain at the required wavelength, this photodetector can be operated between approximately 350 and 1100 nm (where the responsivity has a steep fall from its peak). Alternatively, a combination of photodetectors can be used and the responsivity increased over a larger bandwidth. A combination of Si and InGaAs photodetectors can be used to increase the bandwidth of light detection, therefore across a wider range of wavelengths. These types of systems typically cascade photodetectors on top of each other, with the lower-

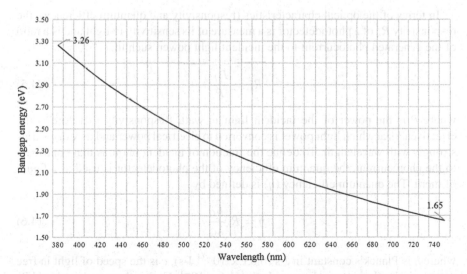

Fig. 4.4 Ideal range of photodetector material energy bandgaps in eV to effectively detect incident visible light (~380–750 nm specifically)

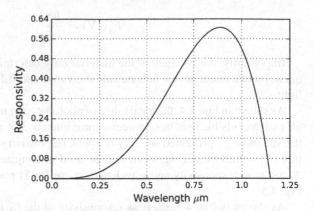

Fig. 4.5 A typical representation of a responsivity (in units mA/W) and wavelength curve for a Si photodetector

wavelength device placed at the bottom. As a result, the top device (in this case the InGaAs photodetector) acts as a window for lower wavelengths; thus incident photons are not converted to electrical energy in this photodetector but can pass through the device towards to the Si photodetector. By combining the electrical outputs, the system is therefore able to detect radiation from visible light up to near-IR. The primary characteristics that determine the performance of a photodetector are its

- responsivity,
- quantum efficiency,
- bandwidth/speed of response,
- dark current, and
- noise-equivalent power (NEP).

In terms of the optical characteristics (responsivity and quantum efficiency), the responsivity R_λ of a photodetector is a measure of its sensitivity to light and is a ratio of the generated photocurrent to the incident light power, such that

$$R_\lambda = \frac{I_{ph}}{P} \tag{4.5}$$

where P is the power of the incident light in watt. Responsivity therefore is the effectiveness at which light power is converted to electrical power and its value varies strongly with wavelength. To define the fraction of incident photons that contribute to the photocurrent, the quantum efficiency of the photodetector is used as a figure of merit. The quantum efficiency (η) is defined by

$$\eta = R_\lambda \frac{hc}{\lambda q} \tag{4.6}$$

where h is Planck's constant in J-s (6.63×10^{-34} J-s), c is the speed of light in free space, q is the elementary electron charge (1.6×10^{-19} C), λ is the wavelength and R_λ

is the responsivity of the photodetector at this wavelength. The quantum efficiency can be simplified to

$$\eta = 1240\left(\frac{R_\lambda}{\lambda}\right) \qquad (4.7)$$

where λ is then specified in nm. Electrically, the photodetector can be completely defined by its bandwidth, dark current, and NEP. The bandwidth of the photodetector, therefore the frequency at which the photodetector output reduces by 3 dB, is indirectly determined by its junction capacitance. The junction capacitance is directly proportionate to the diffused area and inversely proportionate to the depth or thickness (commonly referred to as the width) of the depletion region. The junction capacitance is additionally reliant on the supplied biasing voltage and is given by

$$C_J = \frac{\varepsilon_s\varepsilon_0 A}{W_d} \qquad (4.8)$$

where ε_s and ε_0 are the permittivity of the semiconductor material and free space, respectively, A is the area of the active region and W_d is the width of the depletion region. The dependence on the bias voltage is seen in the calculation of W_d, where

$$W_d = \sqrt{2\varepsilon_s\varepsilon_0\mu\rho(V_A + V_{bi})} \qquad (4.9)$$

where μ is the mobility of the semiconductor material, ρ is the resistivity of the semiconductor material, V_{bi} is the built-in voltage, and V_A is the applied bias voltage. To determine the bandwidth of the photodetector, used in a circuit, the 3 dB frequency (f_{3dB}) is determined by

$$f_{3dB} = \frac{0.35}{\tau_R} \qquad (4.10)$$

where τ_R is the response time in seconds, consisting of three primary factors, namely

- the drift time, which relates the time of charge collection of carriers within the depletion region,
- the diffuse time, which is similar to the drift time, but concerns the collection within the undepleted region, and
- the time constant that depended on the resistance-capacitance (RC) ratio of the photodetector/circuit combination.

As a result, the total value of τ_R is determined by

$$\tau_R = \sqrt{\tau_{drift}^2 + \tau_{diffused}^2 + \tau_{RC}^2} \qquad (4.11)$$

Importantly, the RC time constant depends on a combination of resistances and capacitances. In terms of resistance, the series resistance of the photodetector and the load resistor of the circuit plays a role, and for the capacitance, the junction capacitance C_J and any parasitic capacitances combine to determine the RC time constant. Generally, in relatively small active area photodetectors, the response time is dominated by diffusion time (diffusion time therefore having the largest effect), and for larger active areas, the RC time constant dominates the response time.

The dark current of a photodetector is another important electrical characteristic that influence the performance of the device, specifically the $1/f$ noise, and is defined as the current flowing (under reverse bias) when there is no incident light on the photodetector. Dark current should generally be kept as low as possible, and it is a function of the reverse saturation current I_{sat} such that

$$I_{dark} = I_{sat}\left(e^{\frac{qV_A}{k_B T}} - 1\right) \tag{4.12}$$

where T is the absolute temperature of operation in K. Illuminating the photodetector will shift the curve of current versus voltage (also referred to in practice as the I-V curve) by the quantity of generated photocurrent, as a constant value. Finally, the NEP of the system is a measure of the incident power on the photodetector which produces a photocurrent equivalent to the noise current, such that

$$\text{NEP} = \frac{I_{tn}}{R_\lambda} \tag{4.13}$$

where I_{tn} is the total noise current induced in the photodetector, calculated by

$$I_{tn} = \sqrt{I_{sn}^2 + I_{jn}^2} \tag{4.14}$$

where I_{sn} and I_{jn} are the shot noise and Johnson (thermal) noise, respectively.

In Table 4.1, a list of commonly used photodetectors are presented with their accompanying bandgap properties as well as the cut-off wavelength of each material. The list provided in Table 4.1 also shows typical IR photodetector as a reference of materials used for these types of devices. This list is adapted from Lambrechts and Sinha [11] and expanded on to include visible light photodetectors.

As shown in Table 4.1, each material (or combination of materials) has a specific bandgap associated with it, resulting in a cut-off wavelength where its peak responsivity is achieved. In Li-Fi, the photodetector materials that allow detection of visible light is preferred, depending on the wavelength of the LED source. Designing the LED source using the same materials as the photodetector simplifies the optical gain required at the receiver and increases the effectivity of the system. Most commercial LEDs are "*composed of a blue, high-brightness LED with a phosphorous coating that converts blue light into yellow light to simulate the color of incandescent lights*" [6]. This process is the best cost-effective method to yield white light in modern LEDs; however, the frequency response of these lamps is slowed down through the phosphor color-converting procedure. As a result, the maximum achievable data rates of a Li-Fi system is primarily reliant on the grade of the LED source and less on

Table 4.1 Energy bandgap and cut-off wavelength at 300 K of commonly used materials used to manufacture photodetectors—both visible light alternatives (for Li-Fi) and IR alternatives presented, adapted from Lambrechts and Sinha [11]

Material	Symbol	$E_{g,300}$ (eV)	λ_{co} (nm)	Band
Gallium nitride	GaN	3.40	365	Near UV
Zinc oxide	ZnO	3.37	368	Near UV
Indium gallium nitride	$In_xGa_{1-x}N$	0.69–3.40	365–1800	Near UV/IR
Aluminium gallium arsenide	$Al_xGa_{1-x}As$	1.42–2.16	575–875	Visible
Gallium phosphide	GaP	2.26	555	Visible
Indium gallium phosphide	InGaP	1.90	655	Visible
Cadmium telluride	CdTe	1.50	827	Visible
Gallium arsenide	GaAs	1.42	875	Visible
Indium phosphide	InP	1.35	919	Near IR
Silicon	Si	1.10	1128	Near IR
Indium gallium arsenide	$In_xGa_{1-x}As$	0.73–0.47	1700–2600	Short IR
Indium gallium arsenide	$In_{0.53}Ga_{0.47}As$	0.75	1655	Short IR
Germanium	Ge	0.66	1800	Short IR
Indium arsenide	InAs	0.36	3400	Medium IR
Indium antimonite	InSb	0.17	5700	Medium IR

the photodetector, which can operate at sufficiently-high frequencies. According to Haas [6], the data rate capabilities of light sources, from slowest to highest, can be listed as

- "phosphor-coated blue LEDs,
- red-green-blue (RGB) LEDs,
- GaN micro-LEDs, and
- RGB lasers"

however, the cost of manufacturing these sources increases as the speed of operation increases [6]. Using receivers with higher sensitivity can also increase the performance of a Li-Fi system. The following section discusses *p-i-n* photodetectors, an improvement on the *pn*-junction photodetector in terms of increasing its responsivity, typically through adding a lightly-doped intrinsic region in the construction of the photodetector.

4.3.2 *p-i-n Photodetectors*

A *p-i-n* photodetector constructed similarly to a *pn*-junction photodetector but with an additional intrinsic layer between the *p*-type and *n*-type material. This intrinsic

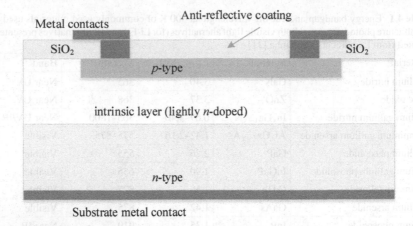

Fig. 4.6 Simplified cross-sectional construction of a *p-i-n* photodetector

layer is relatively wide and undoped region. Figure 4.6 represents the simplified cross-sectional construction of a *p-i-n* photodetector.

The intrinsic region of the device in Fig. 4.6 is where most of the incident photons are absorbed and carriers are generated that contribute to the photocurrent. Effectively, the depletion region between the *p*-type and *n*-type materials is enlarged in a *p-i-n* photodetector and incident photons have a larger likelihood of absorption in this region. The electric field is also large in the intrinsic zone and any electron-hole pairs produced in this region are instantaneously removed by this large electric field and do not have to diffuse out of the region as is the case at the *p*-type and *n*-type regions; additionally, recombination is less likely to occur in the intrinsic region. The wider intrinsic region also leads to a lower capacitance if the device is reverse-biased and this has a positive effect on the speed at which the device can be operated.

To further increase the sensitivity of light detection, as is required to improve performance of Li-Fi systems, avalanche photodetectors (APDs) are typically used. These devices provide a form of built-in first-stage gain of the optical signal and is discussed in the following paragraphs. Avalanche photodetectors provide higher gain of incident light. A single photon striking the receiver produces a cascade of electrons, inherently amplifying the weak light signal significantly.

4.3.3   Avalanche Photodetectors

An APD convert light to electrical energy through a similar mechanism as *pn*-junction or *p-i-n* photodetectors, however, an APD uses avalanche multiplication to provide a relatively high gain of the incident optical signal. These photodetectors are operated at a high reverse bias voltage and in certain scenarios are operated just below its breakdown voltage. In this regime, the charge carriers are excited by absorbed

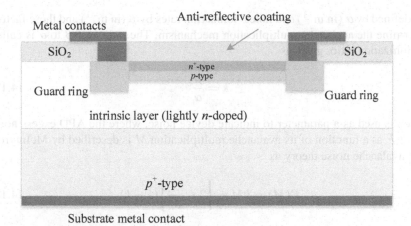

Fig. 4.7 Simplified cross-sectional representation of an avalanche photodetector

photons and strongly accelerated by the internal electric field to generate secondary carriers also referred to as photomultipliers. The generated photocurrent is significantly amplified through the avalanche process leading to sensitive photodetectors requiring lower electrical amplification and therefore subject to lower electronic noise. The avalanche process itself is however subject to quantum and amplification noise and should be monitored as not to offset its low-noise operational advantage. A simplified cross-sectional representation of the APD is presented in Fig. 4.7.

Since the amplification factor of the APD as shown in Fig. 4.7 is strongly dependent on the applied reverse-bias, this amplification factor may vary significantly between devices. Therefore, an APD is typically specified to achieve a specific range of responsivities as a function of a recommended voltage range. An important characteristic of an APD is therefore its avalanche multiplication factor M, a ratio of the multiplied photocurrent I_{phm} to the primary photocurrent I_{ph}, such that

$$M = \frac{I_{phm}}{I_{ph}} \tag{4.15}$$

which is a function of the applied voltage, such that

$$M = \frac{1}{1 - \left(\frac{V_A}{V_{BR}}\right)^m} \tag{4.16}$$

where V_A is the reverse-biased voltage, V_{BR} is the breakdown voltage of the APD, and m is a numerical factor depending on the doping profile of the semiconductor material. For silicon, m is typically between 3 and 6. The number of electron-hole pairs generated during the time that carriers travel a specified distance due to the applied electric field is referred to as the ionization rate. Electron ionization rates are defined by α (in m^{-1}) and holes' ionization rates by β (in m^{-1}) and these factors determine the avalanche multiplication mechanism. The ratio k of β to α is called the ionization ratio, given as

$$k = \frac{\beta}{\alpha} \qquad (4.17)$$

which is used as a parameter to indicate device noise, where the APD excess noise factor F as a function of its avalanche multiplication M is described by McIntyre's [12] avalanche noise theory as

$$F(M) = kM + \left\{2 - \frac{1}{M}\right\}(1 - k) \qquad (4.18)$$

where k is given in (4.17) and is constant. For values of $k < 0.1$ and $M > 20$, McIntyre's avalanche noise theory equation can be simplified to

$$F = 2 + kM. \qquad (4.19)$$

A representation of (4.18) which are the variations of the excess noise factor F with respect to the avalanche multiplication factor M, with constant k, is presented in Fig. 4.8.

As shown in Fig. 4.8, the excess noise factor increases significantly as the avalanche multiplication factor increases, expect for the case where $k = 0$. Additionally, the dark current of an APD is also increased by the avalanche multiplication factor, and is determined by

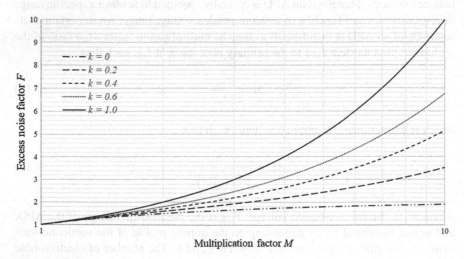

Fig. 4.8 Variations in the excess noise factor as a function of the avalanche multiplication factor [12] in an APD with constant values of k

$$I_D = I_{ds} + I_{db} \times M \tag{4.20}$$

where I_{db} is the bulk leakage current and I_{ds} is the surface leakage current, which limits the sensitivity of the APD under low reverse-bias voltages. Furthermore, the SNR of the APD is also a function of the avalanche multiplication, and is given by

$$\text{SNR} = \frac{I_{ph}^2 M^2}{2q\left(I_{ph} + I_{db}\right)BM^2F + 2qBI_{ds} + \frac{4k_B TB}{R_L}} \tag{4.21}$$

where the last term in the denominator is the thermal noise, q is the elementary electron charge, T is the absolute temperature, R_L is the load resistance, and B is the bandwidth of the measurement.

The responsivity of an APD is relatively crude and therefore not practical for precise measurements, where the p-i-n photodetector is more applicable. However, in practical applications such as Li-Fi where only the presence of light is detected, as opposed to its intensity, APDs are ideal. Silicon-based APDs are typically sensitive at wavelengths between approximately 450 and 1100 nm and maximum responsivity occurs between around 600 and 800 nm, still within the visible spectrum of light. To increase the intrinsic gain of a photodetector, an additional p-type epilayer is added to the APD, in a device that is operated near its breakdown voltage. In a Li-Fi receiver, the incident light is modulated based on its intensity and a device that is capable distinguishing between information-carrying photons with a high sensitivity is a preferred solution in Li-Fi applications.

4.3.4 Single Photon Avalanche Photodetectors

If an APD is operated at a negative bias beyond its breakdown voltage, also referred to as the Geiger-mode, it operates as a single photon avalanche detector (SPAD). The cross-sectional construction is similar to that of the APD and is shown in Fig. 4.9.

As shown in Fig. 4.9, the SPAD is found in the p-type epitaxial level of a CMOS structure and includes two specific layers inherent to SPAD design. These layers are the n-SPAD and the p-SPAD with isolation amid contiguous SPADs achieved by deep p-well layers, in the form of shallow-trench isolation [24].

If the SPAD shown in Fig. 4.9 is operated beyond its breakdown voltage, the electric field becomes extremely large and a distinct charge carrier introduced into the depletion layer can activate a spontaneous avalanche effect [4]. The avalanche effect will continue up until the circuit is sated by reducing the biasing voltage equal to or below the breakdown voltage or above (in the positive direction) it. The quenching circuit can be either passive or active; where a passive quench is simply a resistor placed in series with the SPAD and active quenching is performed by a digital discriminator. In a Li-Fi application, SPADs have a specific and practical use since the system is not necessarily concerned about the magnitude of the incoming

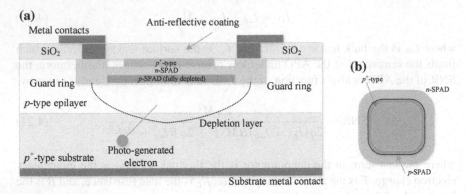

Fig. 4.9 a Cross-sectional representation of a SPAD photodetector operated at a reverse-bias beyond its breakdown voltage and **b** top view of the SPAD active area

pulse but rather about the presence or the absence of this pulse. The high sensitivity and time resolution of a SPAD have been identified as photon counting receivers for Li-Fi systems [24]. Some of its advantageous properties include the power efficiency over extended periods of time, the receivers are typically highly sensitive and capable of approaching quantum-limited sensitivity in exposure to small and weak optical signals, particularly useful for long distance communications [24].

The incident photons on a photodetector convert light energy to electrical energy, as a small variation in current through the applied electric field. However, for this change in electrical signal to be useful, an electrical amplifier is required to amplify the signal to a usable magnitude. Photodetectors typically make use of an amplifier capable of converting the variations in current to a variation in voltage. This amplification process is reviewed in the following section.

4.4   Amplifying Light Signals

Photodetectors are therefore the de facto component to transform incoming pulses of optical light, modulated as information, to electrical signals. The output (current) of these sensors are however very small, and not practical to be used as-is for signal processing. The photons from the light incident on the active area of the photodetector should be transformed to a usable signal. Ordinarily, a TIA is implemented to alter the induced photocurrent I_{ph} to a quantifiable and low-noise output voltage. Such a circuit should have specific characteristics to distinguish between informational light signals and ambient noise, from solar irradiance for example. A complete analysis on TIAs is given in Lambrechts and Sinha [11] and the fundamentals of its operation is presented in this chapter. This section reviews the relevant characteristics of the TIA with respect to its application in a Li-Fi system. The parameters that are of

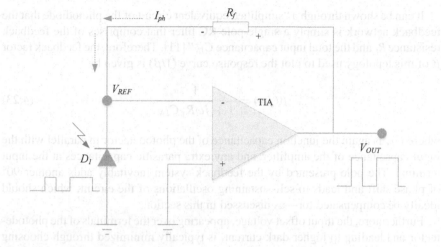

Fig. 4.10 A simplified schematic representation of a TIA utilizing an op-amp to amplify a weak input current to a larger usable voltage

importance to amplify the small current generated by incident photons to a usable voltage are also highlighted.

The primary goal of a TIA is to amplify a weak input current to a large usable voltage. A basic schematic illustration of a TIA is offered in Fig. 4.10. This circuit utilizes an operational amplifier (op-amp) as its active (gain) component and relies on feedback to the inverting-input to provide gain to the input current.

In the TIA representation in Fig. 4.10, the input current as perceived by the inverting input of the op-amp is generated by the active area of a photodetector, D_1. The photodetector typically operates in photovoltaic mode (therefore with no bias voltage applied) and the op-amp ensures that the voltage over D_1 is at zero volts. The photocurrent, I_{ph}, therefore follows the path beginning at the output node of the op-amp, flowing through a feedback resistor R_f, through D_1 and towards the ground node. The measured output voltage, at the output of the op-amp, is simply given as

$$V_{OUT} = I_{ph} \times R_f \qquad (4.22)$$

and this voltage varies in proportion to the intensity of the photons of the light that is incident on the active area of the photodetector. Importantly, the photocurrent is not exactly zero, even if there is no light incident on the photodetector. The generated current with no light present is termed dark current or leakage current and is typically minimized during the construction of the photodiode and careful specification of operating conditions by the manufacturer. The magnitude of this current will also increase as the negative-bias across the photodetector is increased. In various application the photodetector is operated in photoconductive mode (negatively biased) to increase its sensitivity but at the expense of operating speed.

It can be shown through a "simplified equivalent circuit of the photodiode that the feedback network is simply a single-pole RC filter that comprises of the feedback resistance R_f and the total input capacitance C_{IN}" [11]. Therefore, the feedback factor β of this topology used to plot the response curve $(1/\beta)$ is given by

$$\beta(j\omega) = \frac{1}{1 + j\omega R_f C_{IN}} \tag{4.23}$$

where C_{IN} is again the junction capacitance of the photodetector in parallel with the input capacitance of the amplifier, and any extra parasitic capacitances at the input terminal. The pole presented by the feedback system inevitably adds another $90°$ of phase shift and leads to self-sustaining oscillations of the circuit, which should ideally be compensated for—as discussed in this section.

Furthermore, the input offset voltage, appearing over the terminals of the photodetector and leading to higher dark current, is typically minimized through choosing an op-amp with high input resistance, to avoid any current flowing through the op-amp. A high input offset voltage in the system leads to lower dynamic range and ultimately more noise in the system, effectively decreasing the detection range of the photodetector. In a Li-Fi system, this would mean that the distance between the user and the light source decreases and higher-power light sources are required, a typical scenario which is best to be avoided and rectified during the design phase. The ultimate speed at which the entire system can operate determines the attainable data rate in a VLC, assuming that the light source is capable of turning on and off at a comparable frequency. High-speed op-amp TIA implementations for optical signal amplification are therefore dependent on three primary design trade-offs,

- signal bandwidth,
- closed-loop bandwidth (consisting of the open-loop bandwidth, gain, and the total input capacitance to the inverting input of the op-amp), and
- equivalent noise bandwidth.

The total input capacitance of the closed-loop system additionally depends on

- the capacitance of the photodetector as a function of the physical size of its active area and the quality of the insulating oxide between layers,
- the common-mode and differential mode capacitance of the op-amp input terminals, and
- any extra parasitic capacitance introduced into the network such as metal tracks—especially apparent and considerable at high-frequency operation.

A first-order estimate of the maximum likely closed-loop bandwidth that results in a stable circuit (therefore does not oscillate and having a $45°$ phase margin) can be calculated by

$$f_{45} = \sqrt{\frac{f_u}{2\pi R_f (C_{IN} + C_M + C_D)}} \tag{4.24}$$

where f_u is the unity-gain frequency of the op-amp, C_{IN} is the combination of the photodetector capacitance, feedback capacitance (if implemented), and any other parasitic capacitances apparent at the input terminal, and C_M and C_D are the common-mode and differential-mode capacitances of the op-amp, respectively. For larger photodetectors it is generally safe to assume that the C_{IN} is dominated by the capacitance of the photodetector and that the input capacitance of the op-amp can be neglected, however in small photodetectors of sub-100 μm dimensions, these capacitances can become comparable.

Noise bandwidth is a parameter of importance in a VLC system since the influence of noise has a critical part in defining the range and reliability of the link. There are two primary dominant contributors to output noise in a typical TIA configuration, the input voltage noise and thermal noise generated in the feedback resistor R_f (also referred to as Johnson or Nyquist noise). Increasing the resistance of R_f (to ultimately increase the gain of the circuit) inevitably intensifies the thermal noise in the system. Although this noise is not amplified to the output of the TIA it acts as a high noise source to consequent stages. To its advantage, the thermal noise voltage will only grow as function of the square root of the magnitude of the resistance, as proposed by

$$\bar{v}_n^2 = 4k_B T R \qquad (4.25)$$

where k_B is Boltzmann's constant (1.38×10^{-23} J/K), T is the temperature in Kelvin, and R is the value of the resistance. As a result, in high-gain applications, an increase in the feedback resistor is typically preferred as opposed to cascading TIAs, as a cascade configuration would amplify each input-referred thermal noise source from the previous circuit. A more reliable circuit, compared to a large feedback circuit or a cascaded topology, is to have a relatively large gain factor (feedback resistor) to amplify the photocurrent to a usable voltage; followed by a pure voltage amplification step. This ensures a lower input-referred thermal noise source at the voltage amplifier from the feedback resistor used by the TIA.

Furthermore, for additional stability a compensation capacitor C_f in the feedback network of the TIA compensates for gain peaking (phase margin compensation). Figure 4.11 represents the TIA with the compensation capacitor included in the feedback loop.

The feedback capacitor C_f in Fig. 4.11 in parallel with the feedback resistor R_f provides the requisite compensation to guarantee sufficient phase margin. The value of C_f is calculated in relation to the feedback resistance and the total input capacitance of the circuit. This is seen by calculating the feedback factor of the circuit presented in Fig. 4.11, such that

$$\beta(j\omega) = \frac{1 + j\omega R_f C_f}{1 + j\omega R_f (C_f + C_{IN})} \qquad (4.26)$$

which modifies the pole of the original circuit Eq. (4.23) and introduces "a zero in the feedback factor. The zero compensates for the phase shift introduced by the

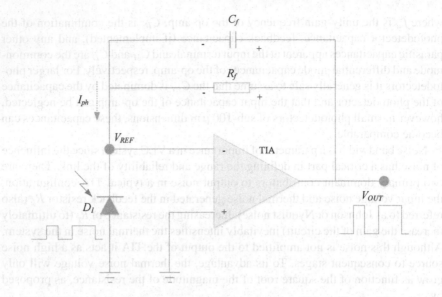

Fig. 4.11 A TIA utilizing a compensation capacitor for stability within the feedback loop

feedback network" [11] but choosing a large feedback capacitor (overcompensation) will inherently reduce the bandwidth of the TIA. Although not necessarily an issue on low-frequency operation, in high-frequency applications such as Li-Fi, where high data rates are to be achieved, bandwidth should be maximized. In these situations, the minimum value of C_f is typically determined that would eliminate self-sustained oscillations and minimize ringing in the circuit. In practice, a slight overcompensation through C_f is typically applied to ensure stable circuits without excessive limitations in bandwidth. It is shown in Bhat [1] that the value of C_f is determined by

$$C_f = \frac{1}{4\pi R_f f_{gbwp}} \left(1 + \sqrt{\left(1 + 8\pi R_f C_{IN} f_{gbwp} \right)} \right) \qquad (4.27)$$

where f_{gbwp} is the unit-gain bandwidth of the op-amp. According to Bhat [1], this calculation of C_f is valid for both small- and large-area photodetectors. If a compensation capacitor is implemented in the feedback network, the closed-loop operational bandwidth of the circuit structure after gain peaking compensation is calculated by

$$f_{3dB} = \frac{1}{2\pi R_f C_f} \qquad (4.28)$$

where f_{3dB} is the subsequent closed-loop operational bandwidth specified in Hz. The corresponding noise-bandwidth (ENBW) of a single pole network is a derivative of the closed-loop bandwidth such that

$$\text{ENBW} = \frac{\pi}{2} f_{3dB} \qquad (4.29)$$

and can be applied to the find the root-mean-square noise due to the resistor in the feedback network and the op-amp current noise.

For an optimal photodetector TIA system, the following considerations should be taken into account; especially when used in a high data rate Li-Fi environment where system integrity and performance is dependent on the limitation incurred by the transmitter-receiver combination in an optoelectronic topology, as is typically employed. Overall, the system bandwidth can be enhanced as well as the total summation of the noise contributions decreased by implementing these guidelines. Further details on these and additional guidelines are discussed in Lambrechts and Sinha [11].

4.5 Considerations in Light Amplification for Li-Fi

This section reviews the important considerations when amplifying light signals generated in a photodetector in a Li-Fi system [3, 5, 9, 13, 15, 17–23, 26, 28]. These parameters include the junction capacitance of the photodetector, the active area of the photodetector, as well as the design parameters of the TIA.

4.5.1 Low Photodetector Junction Capacitance

Firstly, the junction capacitance of the photodetector should be retained at a minimum, or at least as low as manufacturing processes permit. The photodetector junction capacitance as well as the amplifier feedback resistor (and capacitor) form a supplementary noise-gain zero or feedback pole. If the photodetector is bought off-the-shelf, it is recommended to investigate alternatives with low junction capacitance. If a specific size (geometry) of the active area is required, variations in junction capacitance are still noticeable due to differences in manufacturing process and quality of materials used to construct the photodetector.

4.5.2 Small Photodetector Active Area (and Using Optical Gain)

Secondly, concerning the area of the active area, this has a significant effect on the junction capacitance where a smaller photodetector typically has a lower junction capacitance at the cost of lower sensitivity. A smaller photodetector will also present a larger shunt resistance, which is an advantage since it increases the SNR of the integrated system by lowering the noise generated by thermal carriers. This is seen by considering the current generated at the junction of the photodetector, given as

$$I_{jn} = \sqrt{\frac{4k_B T \Delta f}{R_{sh}}}$$ (4.30)

where k_B is again Boltzmann's constant, R_{sh} is the value of the shunt resistance of the photodetector, and Δf is the noise measurement bandwidth. Notice that R_{sh} is located in the denominator, therefore decreasing the generated junction current as it increases (inversely proportional to the size of the active area). A practical guideline is to opt for a smaller photodetector to detect incident light and use external optical gain components such as mirrors or lenses to optimize system performance.

4.5.3 Large TIA Feedback Resistor and Compensation Capacitors

Thirdly, as mentioned earlier in this section, another practical guideline is to use a large feedback resistor in the TIA feedback loop in contrast to a cascaded circuit topology. This avoids input referred thermal noise from individual circuits to be amplified by consequent amplifier stages. Alternatively, a low-to-medium value for the feedback resistor should be used followed by a pure voltage-amplification stage, although the SNR of the input signal (to the voltage amplifier) must be high enough to ensure integrity of the signal. The feedback resistor does conversely lower the response interval of the circuit as the response is subject to both the capacitance at the input C_{IN} and R_f, such that

$$t_R = C_{IN} \times R_f$$ (4.31)

which leads to a longer response time as R_f or C_{IN} is increased. Unstable poles and zeroes in an amplifier design lead to unwanted and spontaneous oscillations and to ensure loop stability, the feedback capacitor as discussed earlier in this section is typically used to ensure circuit stability.

4.5.4 Low Bias Current

A higher sensitivity of the optical receiver can be achieved by lowering the bias current of the op-amp. Op-amps are effectively disposed to present a ratio of leakage current at its input rails with respect to the applied bias current, and as this value (error) increases, the input offset voltage will inherently increase and influence the sensitivity of the photodetector. Input offset voltage additionally fluctuates with temperature and in circuit that is operating in photovoltaic mode, the thermal noise of the feedback resistor may dominate the noise that is present in the entire system and will be enlarged at the op-amp input rail—therefore requiring a low-as-possible bias current to limit input offset voltage variations.

Using a Li-Fi system has certain advantages and limitations, reviewed in this chapter, but a common misconception of Li-Fi is its inability to be operated in daylight due to interference from solar irradiance. The following section briefly discusses solar irradiance in a VLC based on literature presented on the topic, and highlights the primary influences that it may have on Li-Fi transmissions.

4.6   Solar Irradiance in a VLC

The presence of sunlight in a VLC system can have detrimental effects on the optical signal emanating from a LED source and can degrade the SNR, BER, and the overall data rate [8]. In Islim et al. [8], however, solar irradiance is dubbed a misconception surrounding VLC. Islim et al. [8] categorizes solar irradiance (with a constant flux density of 1366.1 W/m^2 spanning wavelengths of 250–2500 nm) in VLC as a large source of shot noise as opposed to an interfering source as the intensity of the sunlight does not differ significantly over short time intervals. As a result, a modulation scheme such as optical frequency division multiplexing (OFDM) can effectively assign the cyphers across the operational frequency subcarriers of the assigned spectrum. Shot noise will inherently reduce the data rate of VLC systems, but the use of an optical band-pass blue-light filter (off-the-shelf) can limit the degradation experienced by solar irradiance [8].

The following section reviews the advantages and limitations of Li-Fi, used in environments with low and high ambient light.

4.7   Advantages and Limitations of Li-Fi

There are various advantages of using Li-Fi as a last mile solution, including its long-term efficiency, high data rates, availability, cost-effectiveness, and its security. These areas are concisely discussed in the subsequent subsections.

4.7.1   Efficiency

One of the driving factors of using Li-Fi is its energy efficiency. The energy efficiency of Li-Fi stems directly from the fact that LED lights are used as the source of transmitting data, and since these light sources are already used to illuminate dark areas, implementing the same LEDs for transmitting data virtually requires no additional power.

4.7.2   Data Rates

Another driving factor of Li-Fi is its potential data rates that can be delivered. Although current practical applications are still limited in data rates, primarily due to the quality and type of LEDs used, using the visible spectrum for data transmission has the potential of data rates that surpass the abilities of Wi-Fi or long-term evolution (LTE).

4.7.3   Availability

Similar to the argument of efficiency, Li-Fi could essentially be available anywhere that a light source is being used to illuminate an area. If the backhaul network is set up and the Li-Fi architecture planned out, additional Li-Fi access points are not expensive or complex to add and expand on the network infrastructure.

4.7.4   Cost-Effective

Apart from its initial investment, maintaining and expanding a Li-Fi network is a cost-effective exercise. Its low-power usage and low-cost source (LED) replacements ensure that even in rural or developing countries, long-term maintenance and expandability can be achieved.

4.7.5   Security

Using light technology to transmit data has additional security-related advantages. Light cannot pass through opaque structures and require LoS and therefore the possibility of intercepted signals or misused transmissions are limited and the security of individual users can be prioritized.

As with any technology, its advantages can often be offset by its inherent limitations, which are often direct trade-offs from its primary advantages. The following paragraphs reviews some of the limitations of Li-Fi that should ideally be overcome for it to become an integral technology in last mile solutions.

4.7.6   Constant Light Source

In order to access data transmissions from a light source, it should be switched on constantly. The efficiency of the system is therefore compromised if a light source is on during a time where illumination is not necessary (during the daytime for example).

4.7.7   LoS

The advantage of LoS communication in terms of security could also be overshadowed and considered a limitation of Li-Fi in terms of practicality. Receiving sensors must be directed at the source to ensure high quality and uninterrupted connections.

4.7.8   Ambient Light/Solar Irradiance

If used in direct sunlight, or if ambient light is directed onto the receiver from nearby sources, there might be a degradation of signal quality or transmission speed, depending on the spurious sources. Additionally, malicious data transmissions can be achieved if an unwanted light source is directed to a receiver that is committed to a different transmission.

4.7.9   High Initial Capital Investment

As with any novel technology that does not have any current infrastructure, the initial investment of rolling out the technology to potential environments is high. In developing countries and rural areas, where the long-term use of this technology would be beneficial, initial investments might have to come from third-party investors since governmental initiatives that support Li-Fi are limited.

4.7.10   In Development

Li-Fi is still in development and the high cost of initial investments and of the hardware required to drive LED sources must be reduced significantly before the technology is mature enough to be rolled out in high volumes. Although research into achieving this is actively pursued, the time to maturity might become a caveat that leads to other technologies overtaking and replacing it. Furthermore, currently

Li-Fi provides only a means for data downlinks, and uplinks are still a challenge. Data and information can therefore be delivered to a user, but with the growing use of cloud computing, uploading information through Li-Fi will become an important consideration in its ultimate success.

To mitigate some of the limitations of Li-Fi and exploit its advantages, networked systems are often deployed as hybrid systems, taking advantage of several technologies to provide seamless and uninterrupted last mile access to users. One popular solution is to use technologies such as LTE and Wi-Fi in conjunction with Li-Fi to create a hybrid solution capable of delivering high-bandwidth data rates in various scenarios. This hybrid solution is reviewed in the following section.

4.8   Hybrid Solutions (Li-Fi/Wi-Fi/LTE)

Li-Fi does not have to operate in isolation and hybrid solutions combining Li-Fi with more traditional communication standards such as Wi-Fi and Global System for Mobile communications (GSM) (LTE) can have multiple benefits in a networking topology. In Fig. 4.12, a simplified representation of a hybrid solution is shown, combining various technologies to achieve seamless and uninterrupted connections in environments that incur different limitations.

As shown in Fig. 4.12, a user should be able to connect seamlessly to a variety of communication protocols depending on the immediate environment and the limitations that may hinder a specific protocol at a given time. The example in Fig. 4.12 shows a user in a residential setting, with communication alternatives based on their position. For example, a central node in the building could receive an Internet connection from a nearby macro- or picocell, enabling 3G or 4G (LTE) communication

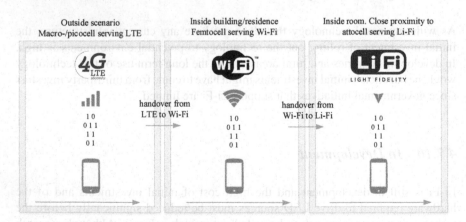

Fig. 4.12   A simplified representation of a hybrid solution implementing Li-Fi, Wi-Fi and GSM (LTE) through a combinations of cells to distribute information

Table 4.2 Comparative parameters among GSM (LTE), Wi-Fi, and Li-Fi to determine the ideal combination in an architecture where geographical limitations often encourage hybrid solutions

Parameter	LTE	Wi-Fi	Li-Fi
Type of spectrum	Radio waves	Radio waves	Visible light
Defining standard	3GPP	IEEE 802.11	IEEE 802.15.7
Range	>1 km	< 300 m	<10 m
Transfer speed	~50 Mbps	<1 Gbps	~1 Gbps
Transceiver placement	nLoS	nLoS	LoS
Bandwidth	Limited	Limited	Virtually unlimited
Power consumption	High	Medium	Low
Cost	High	Low	Initially high/low

when moving in areas where Wi-Fi or Li-Fi is not available. Closer to the building, a Wi-Fi access point from a femtocell distributes wireless Internet from a network connected to a service provider through a copper or fiber cable. The Wi-Fi signal would be distributed throughout the residence to enable Internet-connected devices and various IoT devices to the Internet, considering that the more devices are connected, the more congested the Wi-Fi network would become. Additionally, interference from nearby high-power devices operating at the same frequency (2.4 or 5 GHz) might also have an effect on the integrity of the radio waves emitted from the Wi-Fi access point. In the same scenario, high-speed, virtually unlimited bandwidth, secure, and electromagnetic interference (EMI)-free optical attocells (LED light bulbs) can be used to send and receive data over a Li-Fi network. LoS is required for Li-Fi communication and therefore the architecture of the network would consider this when determining which devices should have access to the Li-Fi network. The optical attocells are programmed for multi-user access, dynamic signal handover, resource sharing, and ensure a heterogeneous network with optimal bandwidth management based on the user requirements. In Table 4.2, a list of comparative parameters between GSM (LTE), Wi-Fi, and Li-Fi is presented, providing a simplified indication of the primary differences between the technologies.

The comparison provided in Table 4.2 gives a proposition of the type of environments where each technology is ideally implemented. Based on the range, transfer speed, bandwidth, power consumption, and cost, a network topology can be created, similar to that of Fig. 4.12, where the advantages of each technology is exploited and the limitations limited. Li-Fi has some clear rewards in relation to transfer speed, power consumption, and long-term cost, but considering the fact that LoS over short distances is required, its usefulness is primarily in areas where devices can be kept relatively stationary and pointed directly to the source. However, in scenarios serving large numbers of people in environments that have limited alternative solutions, especially in rural areas and in developing countries, a multitude of Li-Fi cells (sources) can be a significant and cost-effective solution to provide Internet access to users.

An important consideration when using any technology as a wireless solution to provide networked Internet access, is modeling the channel to determine the propagation characteristics of the signals. However, a single solution does not exist and the channel is dependent on multiple (often virtually infinite) different scenarios. Each scenario is therefore different with respect to the medium, scattering, reflections, transceiver characteristics, and usage. The following section briefly reviews a proposed channel modeling solution presented by Wu et al. [27] as an example of the complexity. It also serves as a starting point when developing a channel model for a unique scenario.

4.9   Li-Fi Channel Modelling

A hybrid system as deliberated in the previous segment combines high-speed data communication that is obtainable by a VLC system with ubiquitous coverage of RF techniques, as described in Wu et al. [27]. A hybrid system can therefore optimize the throughput of a network topology, but it does present certain challenges in its design and choice of access point type and location. Wu et al. [27] proposed a two-stage access point selection (APS) method for a hybrid Li-Fi/Wi-Fi network to achieve optimal throughput and reduce the overall system complexity. The detail of the technique proposed in Wu et al. [27] stands separate to the scope of this book, but the methodology is briefly presented in this section to highlight the critical considerations in such an architecture.

Since the coverage area of Li-Fi is smaller than that of Wi-Fi, Wu et al. [27] proposes a technique that connects all users to the high-bandwidth Li-Fi network, and only the users that experience low throughput at certain times, are handed over to the Wi-Fi network with its superior coverage area. In homogenous networks, a simple APS method that is implemented is to pick the access point that provides the strongest signal to the user. This technique is termed the signal strength strategy (SSS) and is broadly implemented in Wi-Fi topologies. Balancing of the loads is not considered in SSS and a scenario where too many users are connected to a single cell, based on their signal strength requirements, will lead to an unbalanced load and inherently a slowed down network. A similar challenge can exist in a hybrid network, where the larger coverage area of the Wi-Fi could be overloaded when too many Li-Fi devices are experiencing limited throughput due to obstructions, nLoS, or geographical limitations. Wu et al. [27] proposes a variant of fuzzy logic to distribute signals in a hybrid network, basing the choice of connectivity on *degrees of truth* as opposed to the brute *true or false* method.

The system model proposed by Wu et al. [27] is adapted and presented in Fig. 4.13, where a graphic illustration of an indoor hybrid Li-Fi and Wi-Fi architecture is shown, split between four different scenarios (or rooms).

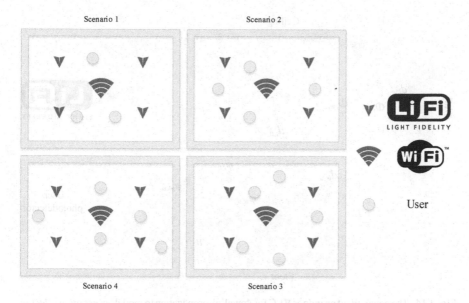

Fig. 4.13 A graphic illustration of an indoor hybrid Li-Fi and Wi-Fi architecture split between four different scenarios, or rooms, adapted from the proposed system model in Wu et al. [27]

As shown in Fig. 4.13, each scenario has a specific number of Li-Fi sources (LED lights that are Li-Fi enabled) as well as a Wi-Fi access point capable of covering the entire room. It is assumed that there is no interference among the Wi-Fi access points, through implementation of carrier sense multiple access with collision detection (CSMA/CA) modulation. The Li-Fi access points use time-division multiple access (TDMA) modulation to facilitate many users, and interference can only occur between Li-Fi access points in the same room since the light cannot penetrate walls of adjacent rooms. The indoor geometry of the proposed model in Wu et al. [27] is adapted and presented in Fig. 4.14. This geometry includes both LoS and nLoS paths and assumes that the Li-Fi access points are fitted to face vertically downwards and a photodetector is placed, facing vertically upwards, to detect incident photons.

From Fig. 4.14, amended from Wu et al. [27], the user (u) and the Li-Fi access point (i), have a direct path between them, corresponding to $d_{i,u}$. The angles of irradiance and incidence associated to the LoS path are designated by $\varphi_{i,u}$ and $\psi_{i,u}$, respectively. In Wu et al. [27] and also in Papanikolaou et al. [14], the LoS Li-Fi channel (H_{LoS}) of is therefore presented by the proposed model, such that

$$H_{LoS}^{i,u} = \frac{(m+1)A_{pd}}{2\pi d_{i,u}^2} \cos(\phi_{i,u}) g_f g_c(\psi_{i,u}) \cos(\psi_{i,u}) \qquad (4.32)$$

where A_{pd} is the physical area of the photodetector, g_f is the optical filter gain, g_c is gain of the optical concentrator, given by

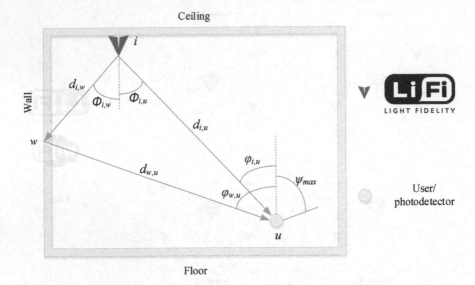

Fig. 4.14 The geometry of an indoor VLC for downlink propagation to model access point selection in a hybrid network topology, adapted from Wu et al. [27]

$$g_c(\psi_{i,u}) = \begin{cases} \frac{n^2}{(\sin(\Psi_{max}))^2}, & 0 \leq \psi_{i,u} \leq \Psi_{max} \\ 0, & \psi_{i,u} > \Psi_{max} \end{cases} \qquad (4.33)$$

where n is the refractive index, and Ψ_{max} is the semi-angle of the field of view of the photodetector. In (4.32),

$$m = \frac{-\ln 2}{\ln(\cos \Phi_{1/2})} \qquad (4.34)$$

is the Lambertian emission order and $\Phi_{1/2}$ is the radiation direction where the intensity is half of the power of the primary ray. For simplicity, Wu et al. [27] only considers first-order reflections in development of the nLoS channel contributions, consisting of the segment from the Li-Fi access point to a point on the wall, w, and the segment between the point on the wall and the user. In Fig. 4.14, these distances are designated by $d_{i,w}$ and $d_{w,u}$ correspondingly. The nLoS channel (H_{nLoS}) is also presented in Wu et al. [27] and the photon current measured by the photodetector, through a combination of the proposed LoS and nLoS channels (resulting in $H_{Li\text{-}Fi}$), is determined through

$$I_{ph} = R_{pd} H_{Li\text{-}Fi}^{i,u} \frac{P_{opt}}{\kappa} \qquad (4.35)$$

where R_{pd} is the responsivity of the photodetector, P_{opt} is the source optical power, and κ is the conversion coefficient between optical and electrical power. Wu et al.

[27] then continues to define the Wi-Fi channels in each scenario and proposes an APS method for the hybrid network, a valuable result in designing such a topology to serve multiple users. As a result, the simulated results of Wu et al. [27] ultimately determines the fairness among users as a percentage of the amount of clients, based on the Li-Fi and Wi-Fi channels and throughput experienced by each user.

Taking into account the optical, electrical, and signal propagation characteristics of a VLC, specifically Li-Fi, the following section highlights certain uses and applications of Li-Fi. As a last mile solution, Li-Fi has a fair number of advantages and these can be exploited in determining its potential applications.

4.10 Li-Fi Applications

This section reviews a number of Li-Fi applications, although there are virtually infinite number of applications that can make use of the technology. As a last mile solution, specifically in rural areas and in developing countries, Li-Fi could alleviate issues where other solutions such as Wi-Fi, GSM, or wired solutions are not viable.

4.10.1 Educational Systems

Smart-class education [25] has become a beneficial innovation for modern schools and universities and allow students to be taught concepts with the aid of technology. In a smart-class, using Li-Fi would enable all students and learners to be collectively connected to a stream of data transmitted from a central source; without bandwidth limitations, and therefore allowing the collective group to experience a wireless interactive learning experience controlled and catered for specifically for each class. Li-Fi would also eliminated network congestion if multiple interactive sessions were conducted at the same time.

4.10.2 Rural Connectivity

Through providing free Internet to rural communities through technologies such as Li-Fi, young people can be encouraged to participate in economic development and develop skills that will be crucial in the already-started fourth industrial revolution. By using low-power LED light sources in conjunction with energy-harvesting through solar panels, light poles that supply a community with free Internet can be maintained and used in schools, youth clubs, or any areas where the technology is distributed to multiple users, even for local businesses. This would enable long-distance learning programs for learners that are not able to travel to distant learning centers, typically by foot and through harsh environments. Furthermore, free Internet through Li-Fi can assist public services such as medical practitioners and teachers to access information that would otherwise not be possible without a constant Internet connection.

4.10.3 Medical Applications

Li-Fi has several advantages when used in the medical field, with various equipment already dependent on Internet connections through Wi-Fi or wired solutions. Medical equipment such as infusion pumps, defibrillators, monitors, lung ventilators, and anesthesia machines require connectivity to send and receive crucial data across networked topologies and ensure that the information is delivered timely and efficiently to physicians and specialists [15]. The drawbacks of using Wi-Fi in these situations are the inherent frequency overlapping and EMI between the medical equipment and the Wi-Fi signals. The limitations are evident in both directions of communication; interference from the medical equipment in the Wi-Fi signals which may temporarily render the communication unusable, as well as interference within the medical equipment from Wi-Fi signals that could render medical test results incorrect or obsolete. Li-Fi in these scenarios would eliminate the frequency overlapping and EMI problems and would ensure constant, high-bandwidth communication between devices.

Additionally, a Li-Fi topology throughout a hospital or clinic would allow medical staff to have access to a patient's vitals at all times without any interruptions from EMI [16]. By fitting the patient with temporary sensors that monitor vitals, this information can be gathered from a Li-Fi source and relayed to the responsible personnel for timely and accurate assessments and treatments.

4.10.4 Transportation

Aircrafts, submarines, trains, and various other transportation vehicles often rely on accurate radio wave signals for communications, security, and geographic information. Traditional radio wave-based information systems for passengers and crews can interfere with these mission critical systems. Such systems are therefore ideally not used throughout an excursion. Li-Fi would eliminate the limitations incurred by radio wave interference and allow major carriers to have mutually exclusive systems for entertainment for passengers and mission critical systems. Overhead lights that are already installed in most airlines and trains could serve as individual in-flight entertainment transmitter that could be used by a passenger at will. Interference between crew and passengers will therefore be limited and there would effectively be no bandwidth limitations for each passenger or crewmember.

The applications of Li-Fi are virtually infinite, and researchers are exploring innovative solutions such as underwater applications, since traditional radio waves transmit poorly under water, disaster management when traditional radio wave communications are not available, traffic management where traffic lights can interactively communicate with traffic, and various other mobile applications where high-bandwidth and secure communications are required. The applications can additionally be modified and catered for developing countries and rural areas with additional limitations that wired solutions and radio waves cannot elucidate. The maturity of Li-Fi and the limitations incurred by the LED sources are therefore crucial to overcome before high volume integration can be achieved.

4.11 Conclusion

This chapter presents a theoretical analysis of Li-Fi as a last mile solution. In its essence, Li-Fi is just another means of serving users with network access, specifically to gain access to the Internet. Li-Fi has the potential to be used as a last mile solution where geographical limitations restricts the use of more traditional technologies. However, since the technology is not yet mature and still under development, this chapter serves to enlighten the reader on the factors that enable this technology.

Li-Fi works on the principle of transmitting information through an optical carrier; therefore, it requires a light source as well as a photodetector capable of receiving these optical signals. These signals can only be used if converted from the optical domain to the electrical domain, and this process is reviewed in this chapter. The photodetector, the device that converts light energy to electrical energy, as well as the TIA that amplifies the signal, are reviewed in detail in this chapter. The fundamental principles of these devices are important to consider when developing a Li-Fi solution, and has a noteworthy consequence on the attainable data rates and coverage area of the VLC system. Furthermore, this chapter reviews more general characteristics of Li-Fi such as the effects that solar irradiance has on the signal and advantages and limitations of the technology. This chapter also discusses potential hybrid solutions to mitigate the limitations of Li-Fi; reviews channel modeling, and presents typical applications of Li-Fi.

References

1. Bhat A (2012) Stabilize your transimpedance amplifier. Maxim Integrated Application Note APP 5129
2. Chacko N, Davies S (2015) Free-space optical networking using the spectrum of visible light. Int J Trends Eng Technol 5(2):217–224
3. Darshith TM, Bhatt C (2017) Prototyping of a Li-Fi communication system. In: 2017 international conference on wireless communications, signal processing and networking (WiSPNET), pp 1804–1807
4. Deng S, Morrison AP (2018) Real-time dark count compensation and temperature monitoring using dual SPADs on the same chip. Electron Lett 54(10):642–643
5. Fakidis J, Videv S, Kucera S, Claussen H, Haas H (2016) Indoor optical wireless power transfer to small cells at nighttime. J Lightwave Technol 34(13):3236–3258

6. Haas H (2017) LiFi is a paradigm-shifting 5G technology. Rev Phys 3:26–31
7. Horecker BL (1943) The absorption spectra of hemoglobin and its derivatives in the visible and near infra-red regions. J Biol Chem 148:173–183
8. Islim MS, Videv S, Safari M, Xie E, McKendry JJD, Gu E, Dawson MD, Haas H (2018) The impact of solar irradiance on visible light communications. J Lightwave Technol 36(12):2376–2386
9. Kalaiselvi VKG, Sangavi A, Dhivya A (2017) Li-Fi technology in traffic light. In: 2017 2nd international conference on computing and communications technologies (ICCCT), pp 404–407
10. Knestrick CL, Cosden TH, Curcio JA (8 Aug 1961) Atmospheric attenuation coefficients in the visible and infrared regions. Naval Research Laboratory: Radiometry Branch. Optics Division
11. Lambrechts JW, Sinha S (2017) SiGe-based re-engineering of electronic warfare subsystems. Springer International Publishing. ISBN: 978-3-319-47403-8 (online)
12. McIntyre RJ (1966) Multiplication noise in uniform avalanche diodes. In: IEEE Transactions on Electron Devices, ED-13, pp 164–168
13. Miranda M, Pradyumna GR (2017) An approach for indoor location estimation to the visually challenged using light fidelity (Li-Fi) technology. In: 2017 international conference on smart technologies for smart nation (SmartTechCon), pp 1055–1058
14. Papanikolaou VK, Bamidis PP, Diamantoulakis PD, Karagiannidis GK (2018) Li-Fi and Wi-Fi with common backhaul: Coordination and resource allocation. In: 2018 IEEE wireless communications and networking conference (WCNC), pp 1–6
15. Poojashree NS, Haripriya P, Muneshwara MS, Anil GN (2014) Li-Fi overview and implementation in medical field. Int J Recent Innov Trends Comput Commun 2(2):288–291
16. Porselvi S, Bhagyalakshmi L, Suman SK (2017) Healthcare monitoring systems using Li-Fi networks. Innovare J Eng Technol 5(2):1–4
17. Romanov OI, Hordashnyk YS, Dong TT (2017) Method for calculating the energy loss of a light signal in a telecommunications Li-Fi system. In: 2017 international conference on information and telecommunication technologies and radio electronics (UkrMiCo), pp 1–7
18. Rosli MA, Ali A, Yahaya NZ (2016) Development of RF energy harvesting technique for Li-Fi application. In: 2016 6th international conference on intelligent and advanced systems (ICIAS), pp 1–6
19. Shamsudheen P, Sureshkumar E, Chunkath J (2016) Performance analysis of visible light communications system for free space optical communication link. Procedia Technol 24:827–833
20. Soltani MD, Safari M, Haas H (2017) On throughput maximization based on optimal update interval in Li-Fi networks. In: 2017 IEEE 28th annual international symposium on personal, indoor, and mobile radio communications, pp 1–6
21. Soni N, Mohta M, Choudhury T (2016) The looming visible light communication Li-Fi: an edge over Wi-Fi. In: 2016 international conference system modeling & advancement in research trends (SMART), pp 201–205
22. Soudgar AI, Kulkarni AU, Surve AR (2017) Li-Fi: an infallible standard for future indoor communication. In: 2017 international conference on electronics, communication and aerospace technology (ICECA), vol 1, pp 393–398
23. Surampudi A, Chapalgaonkar SS, Arumugam P (2018) Can balloons produce Li-Fi? A disaster management perspective. Glob LIFI Congr (GLC) 2018:1–5
24. Takai I, Matsubara H, Soga M, Ohta M, Ogawa M, Yamashita T (2016) Single-photon avalanche diode with enhanced NIR-sensitivity for automotive LIDAR systems. Sensors 16(4):459
25. Tanwar K, Gupta S (2014) Smart class using Li-Fi technology. Int J Eng Sci (IJES) 3(7):16–18
26. Wang Y, Haas H (2015) Dynamic load balancing with handover in hybrid Li-Fi and Wi-Fi networks. J Lightwave Technol 33(22):4671–4682
27. Wu X, Safari M, Haas H (2017) Access point selection for hybrid Li-Fi and Wi-Fi networks. IEEE Trans Commun 65(12):5375–5385
28. Yin L, Haas H (2018) Physical-layer security in multiuser visible light communication networks. IEEE J Sel Areas Commun 36(1):162–174

Chapter 5
Terrestrial and Millimeter-Wave Mobile Backhaul: A Last Mile Solution

Abstract Terrestrial and millimetre-wave (mm-wave) mobile backhaul could be considered as a last mile solution when planning or expanding limited internet infrastructure in emerging markets. These last mile networks offer unprecedented mobile bandwidth capable of serving numerous users at once and are also the basis of future-generation 5G mobile networks. Understanding the principles of mm-wave communications and the benefits and challenges associated with it is important if considering the technology as a mobile backhaul. Its benefits, apart from the advantage of an abundance of unlicensed spectrum, becomes more evident when considering use cases in emerging markets, especially rural areas, where potentially large numbers of users typically group together to obtain internet access. In such a scenario, users are relying on high-capacity access points for a consistently high quality of service. To effectively implement such access areas, high-frequency propagation and attenuation should be reviewed. This chapter researches the theoretical background on mm-wave signal propagation and the difficulties of implementing such high-frequency technologies when compared to more traditional, lower frequency, communications. These principles can be incorporated as a methodological approach to achieve broadband mobile last mile connectivity.

5.1 Introduction

Although the abundant bandwidth available in the millimeter-wave (mm-wave) spectrum is a distinct advantage that could overcome the challenges of providing Gbps data rates to users, there are severe limitations to deal with, an example being the high levels of free-space attenuation. It is however, an economic challenge to provide access to a core network through wired backhaul (e.g. fiber or digital subscriber lines (DSL)) at discrete distance intervals, typically every few hundred meters. A proposed method is to provide users near the repeated (wired) access points access to the network through a large bandwidth wireless backhaul—a heterogeneous deployment as described in Zhang et al. [23]. Deploying such a strategy in mm-wave provides reprieve on the scarcity of signal bandwidth in current cellular networks [4]. However, the limitations of these mm-wave backhaul include

© Springer Nature Switzerland AG 2019
W. Lambrechts and S. Sinha, *Last Mile Internet Access for Emerging Economies*,
Lecture Notes in Networks and Systems 77,
https://doi.org/10.1007/978-3-030-20957-5_5

- high attenuation of the electromagnetic radiation proportional to distance (free-space loss),
- high attenuation and susceptibility to nLoS obstacles such as foliage and walls, and
- limitations and high costs associated with the hardware.

In Hassibi and Hochwald [7] and Du et al. [4] it is noted that even the high bandwidth mm-wave signals can be ineffective and counterproductive if used over larger areas, or in nLoS conditions, and the caveat is that wired connections are also not economically viable [9]. Ideally, radio communications should be delivered as high-speed wireless communication to users, while maintaining efficiency and effectivity of the topology through strategic wireless connections. This essentially implicates combining technologies, wired and wireless, to mitigate limitations and exploit the advantages of each. This book is especially focused on connectivity in rural areas and in emerging markets, since the limitations are from not only a technological perspective, but also have additional and typically, severe financial constraints, and factors such as unskilled workers and little support in governmental policies that further complicate broadband deployment.

Hybrid topologies that use fiber, DSL, mm-wave, and sub-6 GHz communications are capable of optimizing base station (BS) association and from a technological perspective, minimize the mean packet delay in a macro-cell network overlaid with small cells [23]. Low delay times are critical to deliver services with adequate quality of service (QoS). In emerging markets, QoS is of specific importance since network maintenance may be limited to infrequent maintenance schedules by skilled workers (often not situated near the network). Traditional base stations are linked to the core through fiber or DSL links, with low latency and high reliability. Small cells are positioned in locations that are difficult to reach, for example inside lined rooftops or in lampposts. Wired connections to these cells are also not always feasible. High bandwidth wireless signals offer an ideal solution to provide users with networked access; however, as the technology (mm-wave) is maturing, additional limitations and challenges can lead to such projects not being financially viable.

In a typical scenario using wireless repeaters, multiple hops are required to amplify the signal along its intended route and the reliability of these links must be high, as each hop introduces a potential point-of-failure. In multi-hop wireless topologies, sub-6 GHz radio signals have the advantage of nLoS transmission and can lower the complexity of the placement of small cells. Higher frequencies, in the mm-wave range (from 30 to 300 GHz), have bandwidth advantages (as well as beneficial licensing requirements) but suffer from large nLoS attenuation. Hardware capable of processing such high-frequency signals can also become expensive, with reliability also not yet on par with matured technologies.

As an example, Zhang et al. [23] proposes a hybrid network topology that consists of various transmitter and receiver (transceiver) implementations for radio access and a backhaul network, including macro- and small cells, small cell base stations, gateways, and hubs. The locations of the equipment are modelled as independent homogeneous Poisson point processes. The work in Zhang et al. [23] proposes a (optimistic) estimation of the average length of the backhaul and the average number of hops for links between the base station and gateway(s). Such optimization techniques can be employed in planning for last mile access specifically in rural areas where distances between the base stations and the users (subscribers) are relatively large. High-frequency wireless transmission capacity and reliability can also be increased by multiplexing methods for example multiple-input-multiple-output (MIMO) antenna polarization [5, 13]. The fundamental principles used to develop such strategies are reviewed in this chapter and the reader is introduced to the terms and definitions needed to implement optimization techniques such as presented in Zhang et al. [23]. These fundamental principles primarily involve developing channel models for wireless signals that are capable of estimating the power of a transmitted signal at the receiver as a function of the channel distance, obstructions, and wavelength (frequency) of the signal. These models are the focus of this chapter, and it is shown that mm-wave signals, albeit having a capacity advantage, also present cumbersome propagation characteristics. In the following paragraph, a brief introduction into extending the capacity of wireless signals is presented.

5.1.1 Extending Wireless Capacity

There are several fundamental differences between pure mm-wave communications and existing traditional communication systems (for instance Wi-Fi) operating at either 2.4 or 5 GHz. These high bandwidth signals have additional challenges and limitations in the physical layer, medium access control, and routing layers and leads to challenges for newer mm-wave 5G wireless networks [15]. The primary challenges, being

- high levels of propagation loss in free-space,
- high directivity of radio signals,
- weak diffraction abilities,
- increases sensitivity to objects blocking LoS, and
- dynamic requirements due to the mobility of mm-wave signals

require added discernments in the architecture and protocols for reliably deploying mm-wave 5G technology [15]. Essentially, the goal is to increase capacity of wireless channels, and Hemadeh et al. [8] present the three techniques to increase capacity, adapted and presented in Fig. 5.1.

As shown in Fig. 5.1, the three primary techniques that are available to increase the capacity of a wireless signal, are reducing the cell sizes, broadcasting on a new, ideally wider, frequency spectrum, and/or enhancing the signal processing techniques such as modulation and compression. If moving into the realm of the mm-wave spectrum and consideration of the cell size and processing techniques in lower-frequency communications have been exhausted, path-loss modeling becomes crucial, although significantly more complex. However, since the bandwidth advantages of mm-wave offer increased capacity with relatively low complexity in terms of modifying or optimizing current processing techniques, modeling of propagation characteristics becomes essential to determine link budget. Several considerations must also be taken into account, either new considerations resulting from the decreased wavelength, or factors that were assumed negligible in lower-frequency communications. In Table 5.1, a range of considerations for mm-wave communications is presented. These considerations are critical in designing and deploying last mile internet access and the performance and reliability of the wireless network depends on these factors.

Given in Table 5.1 are the critical consideration when designing or deploying an mm-wave communications network, for example in a last mile scenario. The environment plays a crucial role, and LoS and nLoS communications have vastly different requirements, especially true for mm-wave signals that experience high attenuation

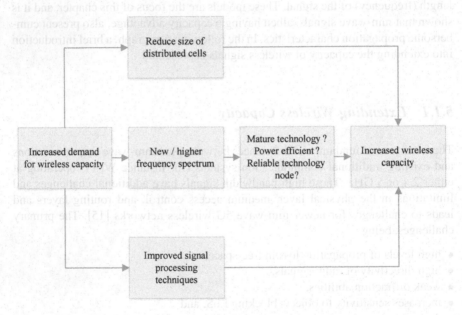

Fig. 5.1 The three primary techniques to increase the capacity of a wireless communication signal, adapted from Hemadeh et al. [8]

Table 5.1 Critical characteristics and consideration in mm-wave propagation channel modelling

Environment	Characteristics	Scenario	Modelling method	Frequency range	Channel measurement
LoS	Temporal	Indoor	Map-based	<100 GHz	Parameter extraction
nLoS	Spatial	Outdoor	Stochastic	>100 GHz	Channel sounding
		Outdoor-indoor	Ray-tracing		
		Vehicular	Geometric		

from physical obstructions. Furthermore, factors such as temporal (time variant) or spatial (position in space) characteristics, indoor, outdoor, or vehicular scenarios, map-based, stochastic, ray-tracing, or geometric modelling, parameter extraction versus channel sounding, and the frequency range (above or below 100 GHz) are critical to consider for mm-wave channel modelling.

This chapter primarily aims to identify propagation path-loss models for challenging wireless communications used as backhaul and reviews fundamental principles and techniques that can be used to predict the link budget in network architectures capable of serving multiple users in rural communities or in urban environments. The goal is therefore to summarize the applicable models and provide researchers with technical references to each of these models.

The first review describes the general characteristics of wireless signal propagation and lists the parameters that influence signal integrity between the transmitter and the receiver. The section is followed by a general review on the principals of path-loss modelling which includes defining and reviewing free-space loss mechanisms. The applicable propagation models for lower (below 11 GHz) frequencies (compared to mm-wave) followed by models developed for frequency communications above 11 GHz and well into the mm-wave spectrum. Since this chapter has already referenced the term backhaul several times already, the following section briefly describes this term by comparing it to fronthaul, both terms often used in communications architecture.

5.2 Backhaul and Fronthaul

In Fig. 5.2, the basic differences between a mobile backhaul and mobile fronthaul are outlined.

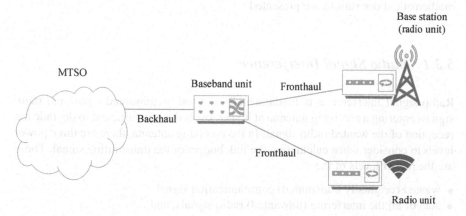

Fig. 5.2 Architectural comparison between mobile backhaul and fronthaul

In its simplest form, a mobile backhaul is required to connect the mobile network to the wired (core) network by backhauling traffic from geographically dispersed cells to mobile switching telephone offices (MTSOs). In a typical macro-cell site, a baseband unit is connected to a radio unit (Fig. 5.2). The baseband unit is expected to process user data as well as control data, whereas the radio unit generates the wireless radio signals that are transmitted across the airwaves. The mobile backhaul therefore forms part of the transport network, is dynamic, and has low latency (offering data rates in Mbps). A mobile fronthaul is specifically related to centralized baseband controllers as well as standalone radio heads, installed at remote location often several kilometers away from the radio access network. All of the equipment is therefore physically located a fair distance away from the mobile backhaul. The radio unit in the mobile fronthaul is often referred to as the remote radio head. The baseband unit is centralized and serves multiple radio units (including larger base stations as seen in Fig. 5.2). Optical interconnects are used between the centralized baseband unit and the multiple radio units, and this architecture is referred to as mobile fronthaul. The mobile fronthaul is static and has very low latency (offering data rates in Gbps).

5.3 Wireless Signal Propagation Challenges

This section highlights some of the unique challenges faced when designing an mm-wave communications system. The unique traits of mm-wave signals make accurate modeling of path-loss models complex and cumbersome, however, many models rely on (albeit limited [8]) measured data. The complexity of mm-wave modelling is primarily due to high levels of radio signal interference, large effects of the propagation mechanisms of free-space, gaseous attenuation, diffraction and scattering, tropospheric and ionosphere attenuation, physical objects, multi-path effects, and shadowing. Each of these challenges are briefly reviewed in this section, before the mathematical derivations are presented.

5.3.1 Radio Signal Interference

Radio signal interference is fundamentally defined as unwanted (spurious) radio signals entering a receiving antenna at power levels that are sufficient to degrade the reception of the wanted radio signal. In the receiving antenna, there are three power levels to consider when calculating the link budget of the transmitting signal. These are the power levels of the

- wanted (originally transmitted) communication signal,
- sum of all the interfering (unwanted) radio signals, and
- noise generated in the system.

There exists a minimum power level for the wanted signal that must be higher than the power level and the sum of the interfering signals, referred to as the signal-to-noise ratio, a typical design-requirement in transceivers. The convention of this ratio is given as

$$10 \log \frac{C}{N + \sum I} \tag{5.1}$$

where C is the power of the wanted signal, N represents the noise power, and I is the power of the intrusive signals. The ratio is expressed in dB and gives the relative strengths of the signals required to calculate interference in a system, therefore the link budget. In high-frequency mm-wave systems, the noise power is typically dominated by circuit noise and external factors such as oxygen absorption (particularly in the 60 GHz band). Furthermore, the strength of a radio signal varies significantly, even when the transmitting and receiving antennas are stationary.

5.3.2 Free-Space Loss

Free-space path loss (FSPL) (or spreading loss) is a fundamental physical phenomena occurring when a signal is transmitted through free-space (air) and is quantified by the Friis transmission equation

$$\text{FSPL} = \left(\frac{4 \pi d}{\lambda} \right)^2 \tag{5.2}$$

where d is the distance between the source and the receiving antenna and λ is the wavelength of the signal (therefore its frequency). This well-known equation shows the square-relationship that is inherent to all electromagnetic waves and the attenuation increases with increasing distance and decreasing wavelength (increased frequency). As will become apparent in this chapter, the inverse relationship to frequency becomes a severe challenge for mm-wave backhaul networks, requiring either high-power transmitting antennas or frequently repeated amplification of the signal.

5.3.3 Gaseous Attenuation

Atmospheric gases (such as oxygen, ozone, precipitation, and water vapor) significantly attenuate high-frequency signals, if the wavelength is comparable to that of the resonant frequency of the gas molecules. These attenuation factors are often specified as constants for specific frequencies (in dB/km) and increase in value as the operating frequency increases. External factors such as pressure, temperature, and humidity also influence the attenuation experienced by the radio signal and are often included in path-loss modelling of high-frequency systems (and typically ignored for lower-frequency communication).

5.3.4 Diffraction

Obstructions in the path of a radio signal lead to diffraction losses of the wanted signal. Along its terrestrial path, a signal may experience numerous obstructions, as a combination of the Earth's curvature (for long-range low-frequency communication), hills, vegetation, or synthetic obstructions. An example of a diffraction mechanism is knife-edge diffraction, and this mechanism is reviewed in Lambrechts and Sinha [12]. Diffraction depends on the relationship between the wavelength of the transmitted signal and the geometric size of the obstruction. Atmospheric refractivity can additionally cause convergence or divergence (de-focusing) of radio signals and further attenuate the wanted signal.

5.3.5 Troposphere/Ionosphere Scattering

Typically used to increase the propagation distance of a radio signal, scattering from the troposphere (ideally for signals of approximately 2 GHz) or ionosphere (ideally for signals below 1 GHz) can cause signal enhancements on terrestrial paths. Although there are significant advantages in increasing propagation distance, these methods are extremely variable and unreliable. Furthermore, mm-wave backhaul communication cannot take advantage of these methods due to the limitations in propagation distance.

5.3.6 Multi-path Propagation

In wireless communications, reflections from objects such as walls and buildings, atmospheric inhomogeneity/ducting, or ground reflections can create simultaneous multiple propagation paths and lead to frequency-selective signal variability. In urban environment, objects such as moving vehicles can also cause variability in the time-domain, further complicating multi-path propagation characteristics. Multipath fading is a result of multipath interference mainly caused by variability in the amplitude as well as the phase of the received signal leading to phase cancellation. Multipath fading consequences include fast fluctuations in signal power over distance or time intervals, arbitrary frequency modulation as a result of fluctuating Doppler shifts for altered multipath signals, and/or time dispersion or resonances produced by multipath propagation delays.

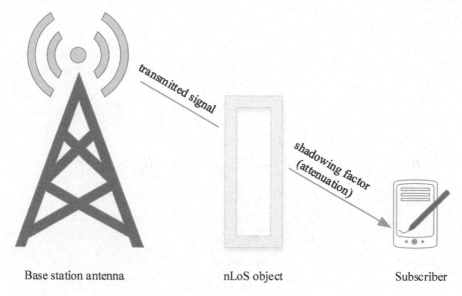

| Base station antenna | nLoS object | Subscriber |

Fig. 5.3 Visual representation of the shadowing factor *s*

5.3.7 Shadowing

Shadowing is the portion of the radio wave that is eradicated due to absorption, reflection, scattering, or diffraction (but not multipath effects or Doppler spread). In Fig. 5.3, the principle of shadowing is visually represented.

As shown in Fig. 5.3, if a base station antenna transmits to a subscriber, an object in the middle of the transmitter and receiver would create a *shadow* on the client antenna, similar to light from a lamppost obstructed by an object. The shadowing factor is a Gaussian dispersed random variable and defined by its standard deviation, typically provided as a constant value in path-loss models.

Many path-loss models approximate the immediate scenario as a street canyon. In the following section, the street canyon approximation is reviewed as it also forms an integral part of path-loss modelling for mm-wave signals. The street canyon approximation also takes into account phenomena such as shadowing, and lends itself to determining the constant attenuation factors in mm-wave path-loss models.

5.4 Street Canyon Approximation

A street canyon, often used in air quality modeling [22], refers to a geometry in an urban environment that resembles a canyon in nature, where a narrow street is lined up continuously with buildings on either side. The term is also used to define larger avenues where both sides of the street are not necessary lined up with buildings at

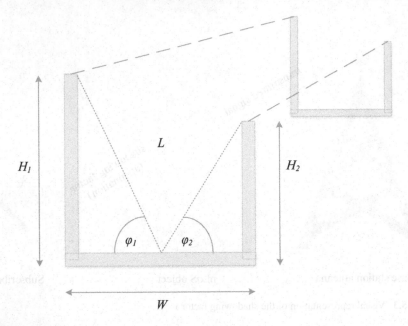

Fig. 5.4 Simplified representation of **a** symmetric and **b** asymmetric street canyon, defining the width (W) and height (H_1 and H_2) of either side of the canyon, and the distance between two intersections (L)

all times, more relatable to common urban environments. Figure 5.4 is a simplified representation of a street canyon.

As shown in Fig. 5.4, the aspect ratio of the canyon and the height of the structures define the street canyon. The geometry of the street canyon is described by this aspect ratio, defined by the height (H) and the width (W) of the *canyon*. Further sub-classifications also exist for street canyons, based on the distance between two major intersections along the street (L). There are several classification of the aspect ratio of a street canyon, summarized in Table 5.2.

As summarized in Table 5.2, a street canyon can therefore be classified as either a regular-, avenue-, or deep street canyon, with sub-classifications of short-, medium-, and long-street canyons based on the distance between intersections. Other classifications include symmetric or asymmetric street canyons based on the variation in height of the buildings. In general, several effects are witnessed in urban street canyons. These include

- unusual and unpredictable variations in wind strength and direction,
- absorption of heat energy resulting in urban heat islands [11],
- higher levels of polluted air not able to clear the vicinity,
- noise pollution and echoing, and
- disruptions in wireless communication signals leading to increased propagation path loss and additional interference from nearby sources or signal reflections.

Table 5.2 Classifications of a street canyon based on the aspect ratio of height (*H*) and width (*W*) of the canyon

Classification	Aspect ratio (*H/W*)
Regular	Approximately 1
Avenue	<0.5
Deep	Approximately 2
Sub-classification	Aspect ratio *L/H*
Short	Approximately 3
Medium	Approximately 5
Long	Approximately 7
Symmetry classification	Description
Symmetric/even	Buildings have comparably similar heights
Asymmetric	Buildings vary significantly in height

Includes sub-classifications based on the aspect ratio *L/H*. Adapted from Vardoulakis et al. [22]

The last characteristic in the list above is what is of particular interest considering the theme of this chapter. Additional path losses of radio signals, particularly in the RF and mm-wave domains, lead to factors that need to be modeled when planning and distributing communication equipment in urban environments. Similar to pollution modeling in urban areas, based on the likeness to street canyons, wireless communications models exist to account for variability in the radio signals. Work such as Rasekh et al. [18] aims to define wireless propagation channels in street canyons for mm-wave (specifically at 60 GHz) signals. In this case, reflection modelling is achieved by considering carrier frequency, although when using Fresnel's reflection equations, coefficients do not vary significantly if the wave number is constant. Rasekh et al. [18] also considers other factors such as surface roughness and its effects on reflection and diffraction of signals. However, even when approximating a street canyon environment based on its aspect ratio, reflection coefficients, surface roughness, and carrier frequency, it must be noted that each scenario will still yield significantly different results in the final model. As a result, various *standard* empirical propagation models are often used to approximate the environment as best as possible. To complicate the modeling of wireless signals in street canyons, many street canyon models also exist. The propagation model should therefore be combined with the street canyon model in order to define an accurate final model.

5.5 Path-Loss Modeling General Approach

This section briefly presents the general approach of constructing path-loss models for radio signals, presented in parts in Chaps. 1 and 3 of this book. The aim of this section is not to study and review the general path-loss model, as this is widely covered in literature, but rather to present the first principles in lieu of the following sections that highlight particular models used based on their frequency characteristics and direct environment. Path-loss propagation models are vital to approximate the link budget of a radio signal, i.e. determine how much power is needed at the transmitter and how sensitive the receiver needs to be. Path-loss is essentially unwanted reductions in signal strength alongside the path it follows between the source and receiving components. It leads to a reduction/attenuation in the power density of the originally transmitted signal as it propagates through a medium (typically free-space). For maturing technologies such as 5G (with candidate carrier frequencies of 15, 28, 38, 60, and 73 GHz), reviews on transmission path-loss models for urban environments have been presented in works such as Kim et al. [10], Sun et al. [21], and Sulyman et al. [20]. The type of propagation model (classification) depends on the strategy employed, and there are typically three primary strategies associated with path-loss modelling. The general classifications of path-loss transmission models are presented in Fig. 5.5.

As shown in Fig. 5.5, the general classifications of path-loss include three primary categories: deterministic, statistical (stochastic), or empirical. These three classifications define the way that the data of a specific scenario is gathered or generated to predict the signal strength at the receiver. Deterministic methods are based on the theoretical and physical laws that govern electromagnetic radiation. Computational power of these models is often intense, as they require large amounts of data to define the propagation model (such as complete three-dimensional maps of the environments). Maxwell's equations are used to define these propagation models and predictions are purely based on how accurate the immediate environment can be modeled. Ray-tracing models (representing wave fronts as particles) are typical examples of deterministic path-loss models, a technique that is also used in computer generated light-wave modeling, requiring powerful processors. Statistical or stochastic models are based on random variables, finding the probability density

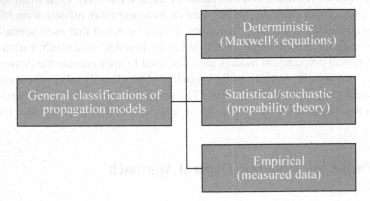

Fig. 5.5 The general classifications of path-loss propagation models for wireless signals: deterministic, statistical, and empirical

function of a transmission in a particular scenario and employing statistical models to predict signal strength. These models are typically the least precise, but, offer solutions that necessitate the lowest amount of computational power. The accuracy of the stochastic models improves with higher frequency, essentially because of the high attenuation as a function of distance and the lower complexity of the immediate environment. Finally, empirical models are based solely on observed and practical data and are used most-often in conjunction with the fundamental equations of electromagnetic propagation to forecast path-loss in similar surroundings. These models are either non-time-dispersive (such as the Stanford University Interim (SUI) model) or time-dispersive (COST-231 Hata model). Field data for urban, suburban, and rural environments can be used to construct accurate models for these environments, although exact attenuation will differ in each scenario, these models provide the best trade-off in complexity and accuracy.

As an example of an empirical/deterministic scenario approximation for mm-wave path-loss modeling, Du et al. [4] propose the use of relays in mm-wave bands that are capable of supporting high-bandwidth data transfers at traditional cellular coverage ranges. These coverage ranges are typically around 1 km, deployed every ~100 m apart, serving consumers in their area and backhauled to the fixed/wired base station using wireless communication. Du et al. [4] present three distinct scenarios of wireless backhaul, a single hop, nearest neighbor, and full connectivity strategy, adapted and presented in Fig. 5.6.

In Fig. 5.6, wireless links among relays are achieved through high-gain directional antennas and optimal design topologies and allocation of bandwidth per user can be deterministically calculated, simulated, empirically measured, and generally modeled. Du et al. [4] also employs the street canyon approximation to propose a scheme of a wireless backhaul that supports Gbps consumer speeds at the brink/edge of the cell. Each user is assigned to a base station and it is assumed that the space among each user and the accompanied relay/base station is equal to the circular coverage of the small cell. For multiple subscribers, the bandwidth and power are split to accommodate the data rate requirements of each user. Since bandwidth in mm-wave is plentiful, the power is predominantly scaled to accommodate multiple users, and it is furthered assumed that all backhaul connections are orthogonal in bandwidth [4]. The path-loss models used in Du et al. [4] are the blocked LoS, the 5GCM Urban Macro (UMa) nLoS, and the 3GPP Urban Micro (UMi) nLoS models, all capable of modeling frequencies upwards of 11 GHz and reviewed in this chapter. Du et al. [4] concludes that by using the street canyon approach and the proposed path-loss models, a linear system positioned along a street at 28 GHz, a 1 Gbps data rate (per user) "inside a cell range of 1 km can be achieved, at 1.5 GHz bandwidth and high-gain single-polarized antennas" [4]. These are promising results for mm-wave backhaul networks deployed as last mile solutions and with maturing technology, a feasible alternative to fully wired topologies.

To evaluate the path-loss models used in works such as Du et al. [4] for mm-wave last mile backhaul, the following section reviews the principles in developing a path-loss propagation model. The general approach essentially considers inherent free-space losses and particular models are expanded upon based on environmental conditions, frequency interaction with the environment, and objects blocking LoS.

5.6   Methodology of a Path-Loss Model

The methodology to construct path-loss propagation models is based on a least-square regression analysis [2, 14, 19, 20] deriving the general path-loss (*PL*) model, expressed in dB, such that

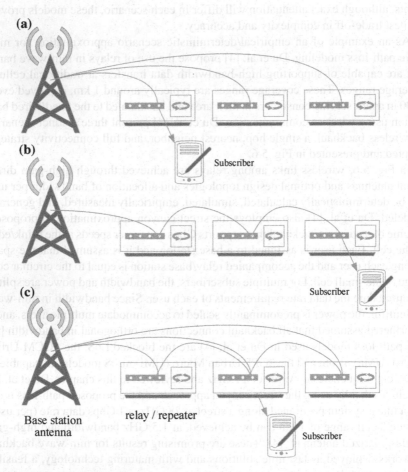

Fig. 5.6 Three wireless backhaul network topologies utilizing relays deployed along points of interest and backhauled to a wired base station coupled to the core network, adapted from Du et al. [4]. The three strategies presented are **a** single hop, **b** nearest neighbor, and **c** full connectivity

$$\overline{PL}(d)[dB] = PL(d_0) + 10n \log_{10}\left(\frac{d}{d_0}\right) \tag{5.3}$$

where d_0 is the orientation node at a 1 km distance, d is the distance between the transmitting source and the receiving node, and n is the path-loss exponent as defined in Chap. 3 of this book. To demonstrate the dependency on frequency (5.3) is typically rewritten as

$$\overline{PL}(d)[dB] = 20 \log_{10}\left(\frac{4\pi d_0}{\lambda}\right) + 10n \log_{10}\left(\frac{d}{d_0}\right) \tag{5.4}$$

where λ is the wavelength of the radio signal. Note in the first term in (5.4), as λ decreases (therefore frequency increases), the path-loss exponentially increases (leading to *higher losses*). Furthermore, nLoS components are additionally added to the path loss as constants, not considering reflections or refraction of the signal in the general model, such that

$$\overline{PL}(d)[dB] = 20 \log_{10}\left(\frac{4\pi d_0}{\lambda}\right) + 10n \log_{10}\left(\frac{d}{d_0}\right) + \sum_i AF_i \tag{5.5}$$

where AF_i is the reduction in dB of the electromagnetic signal as a result of an obtrusive entity that leads to breaking the LoS. Each individual attenuation factor AF_i is described as a function of thickness, concentration, as well as physical size; and practical (empirical) data can be applied for these values. The amount of attenuation of each object is also a function of the carrier wavelength. The link budget of the system, also presented in Chap. 1 of this book, is therefore determined by

$$P_R = P_T - L_T + G_T + G_R - L_{FS} - L_R \tag{5.6}$$

where P_R is the power received by the receiver, for example the strength of the network signal received by the device carried by the user. This received power is a function of various parameters and characteristics within the communication system, including the power/energy of the source signal, P_T, and the losses associated with the transmitter. These losses are internal to the design and the components used by the transmitter and can be experienced where connectors are present that are not properly matched, or any other losses in electronic components due to tolerances or improper design techniques. G_T and G_R in (5.6) represent the gains of the transceiver antennas, respectively. It is therefore typical to see high-gain antennas used in communication networks to ensure that the signal leaving the transmitter experiences a high level of amplification, and the receiver is capable of distinguishing small signals from noise and amplifying these to usable levels before in situ circuit amplification occurs. Furthermore, losses within the medium (channel) should be accounted for, typically referred to free-space losses (L_{FS}), a wavelength/frequency-dependent factor, as well as losses within the receiver, L_R, again through mismatches and sub-optimal circuit design.

The following section reviews fixed wireless access path-loss propagation models for frequencies in the lower frequency spectrum (when compared to mm-wave frequencies), predominantly below 30 GHz. These models act as a precursor to higher frequency (30 up to 300 GHz) models that have additional complex losses associated with its short wavelength.

5.7 Fixed Wireless Access Path-Loss Models (RF Below 30 GHz)

Several empirical propagation models for fixed wireless communications have been proposed, and validated in works such as Abhayawardhana et al. [2]. These propagation models, being frequency-dependent, are typically defined for a specific range of frequencies where the predictions are accurate. Outside of these frequency bands the models are considered less effective and could yield incorrect predictions. For RF networks in the frequency collections below 11 GHz (arbitrarily chosen since some of the reviewed models include frequencies up to 11 GHz), numerous models have been validated and are often being used to design and implement wireless communications in urban or rural areas. Communications in the lower frequency tier (below 30 GHz in this chapter) are common, and as a result, bandwidth is being saturated, while increased growth is still experienced worldwide.

In terms of propagation modeling and path-loss estimation, the traditional models are especially relevant when developing new models for higher, mm-wave, frequencies for future high-bandwidth environments capable of serving numerous subscribers. The lower frequency models are therefore used as a baseline to develop mm-wave models. In Table 5.3, the most commonly used path-loss models for traditional communication standards are summarized with a brief description of its fundamental properties.

Table 5.3 Commonly used path-loss propagation models for frequency bands below 11 GHz

Path-loss model	Description	Frequency range (GHz)
Line-of-sight (LoS) free-space model	–	All frequencies
Stanford University Interim (Erceg) model	Developed specifically for WiMAX	1.9–11
COST-231 Hata model	Okumura-Hata model for urban areas	0.15–2
ECC-33 model	Developed for cellular/microwave	0.7–3.5
Irregular Terrain Model (ITM)	Longley-Rice model	0.02–20
Ericsson 9999	Adapted Hata model	0.15–1.9
Egli VHF/UHF	General purpose UHF/VHF model	0.03–1

The path-loss prediction models listed in Table 5.3 show the frequency bands where these models are most accurate. The general LoS free-space model is developed for all frequency bands, and serves as a baseline in developing all path-loss models (since free-space losses are experienced by all electromagnetic signals).

The following sections reviews the path-loss models applicable to frequency bands where broadband communications can be realized, therefore the

- Stanford University Interim (SUI) (Erceg) model,
- COST-231 Hata model, and the
- ECC-33 model.

The lower frequency models (Ericsson 9999 and Egli VHF/UHF) and the general ITM model are not reviewed in this chapter. The first model, the SUI (Erceg) model, is presented in the following section.

5.7.1 Stanford University Interim (SUI)—Erceg Model

The primary goal of collaborations between Stanford University and the IEEE 802.16 group was to develop a channel model for WiMAX application in suburban environments.[1] As a result, the SUI propagation loss model was developed to calculate the median path loss for IEEE 802.16e (mobile broadband wireless access systems) in three categories, namely

- Category A: maximum path-loss experienced in hilly terrains with moderate-to-heavy foliage concentrations.
- Category B: intermediary path-loss in hilly environments with either sparse vegetation, or with heavy vegetation on a predominantly flat terrain.
- Category C: slightest path-loss conditions in predominantly level landscapes with light vegetation or foliage concentrations.

The proposed SUI model is accurate for operating frequencies below the Ku band (11–18 GHz) and is defined particularly for the multipoint microwave distribution system (MMDS) frequency band in the United States of America (2.5–2.7 GHz). The SUI model is ideally applied in scenarios where the

- coverage radius of the individual cells is less than 10 km,
- height of the antenna at the receiving module ranges between 2 and 10 m,
- height of the transmitting source (base station) antenna ranges among 15 and 40 m, and
- cellular coverage of between 80 and 90% is assumed.

Furthermore, the SUI model includes variations: a basic (Erceg) model, a model that applies correction factors, and an extended model. The basic SUI model, proposed for signal frequencies around and below 2 GHz, receiver antenna height of less than 2 m, and suburban environments, is defined as

---

[1] The SUI model is available online at https://www.xirio-online.com/help/en/sui.html.

$$\overline{PL}(d)[dB] = A + B + s \qquad (5.7)$$

where s is a shadowing effect, and

$$A = 20 \log_{10}\left(\frac{4\pi d_0}{\lambda}\right) \qquad (5.8)$$

and

$$B = 10\gamma \log_{10}\left(\frac{d}{d_0}\right) \qquad (5.9)$$

where γ is the model's path-loss exponent [21]. The shadowing effect is represented by a lognormal distributed parameter and typical values for s range between 8.2 and 10.6 dB for this model. It can be noted that the Erceg model closely resembles the basic propagation model presented in (5.6). The SUI model defines its path-loss exponent as

$$\gamma = a - bh_{BS} + \frac{c}{h_{BS}} \qquad (5.10)$$

where h_{BS} is the height of the base station, and the coefficients a, b, and c relies on the grouping (A, B, or C) and are summarized in Table 5.4.

In Fig. 5.7, the distribution of the path-loss exponent as a function of the suburban environment and the height of the transmitting source/base station is presented. The example uses an arbitrary shadowing factor of 0 dB and the coefficients are presented in Table 5.4.

The distribution of the path-loss exponent for Categories A, B, and C of the basic SUI model is specifically defined to be accurate for a base station height of less than 2 m. As seen in Fig. 5.7, all three categories tends to similar values of γ for a base station height above 2 m and therefore becomes less accurate to distinguish among suburban environments. Noticeable from Fig. 5.7 is the fact that Category C presents the largest path-loss exponent and Category A presents the lowest path-loss exponent.

For the SUI to accommodate operating frequencies above 2 GHz as well as base station heights of between approximately 2 and 10 m, the SUI model incorporates correction factors, and as a result, (5.7) is adapted such that

Table 5.4 Theoretical constant values of the coefficients in the basic (Erceg) SUI model that define the path-loss exponent	Coefficient	Category A (highest losses)	Category B (medium loss)	Category C (lowest loss)
	a	4.6	4.0	3.6
	b	0.0075	0.0065	0.0050
	c	12.6	17.1	20.0

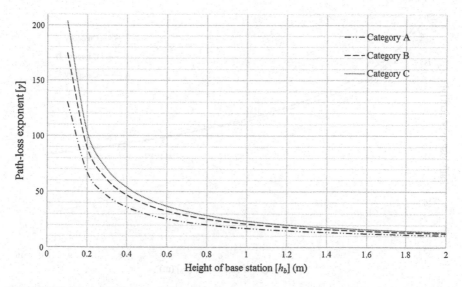

Fig. 5.7 Visual representation of the path-loss exponent of the elementary SUI wireless transmission model showing the variation in category attenuation as a function of height of the base station (from 0 to a maximum of 2 m)

$$\overline{PL}(d)[dB] = A + B + s + \Delta L_{bf} + \Delta L_{bh} \qquad (5.11)$$

where ΔL_{bf} is a frequency correction factor defined as

$$\Delta L_{bf} = 6.0 \log \frac{f}{2000} \qquad (5.12)$$

and ΔL_{bh} is a correction factor concerning the receiver antenna height described by

$$\Delta L_{bh} = -10.8 \log \frac{h}{2} \qquad (5.13)$$

for Category A and B type terrain and

$$\Delta L_{bh} = -20 \log \frac{h}{2} \qquad (5.14)$$

for Category C type terrain. Assuming again a shadowing factor of 0 dB and a carrier frequency of 2.5 GHz, Fig. 5.8 shows the SUI model with correction factors applied.

In Fig. 5.8, it is shown that at base station heights above 3 m the attenuation of Categories A and B dominates (the highest path-loss exponent) according to the SUI model using correction factors, clearly a strong function of the base station height. This variation of the model is used most often in propagation modelling based on the SUI model, where the extended model (using a modified correction factor and a

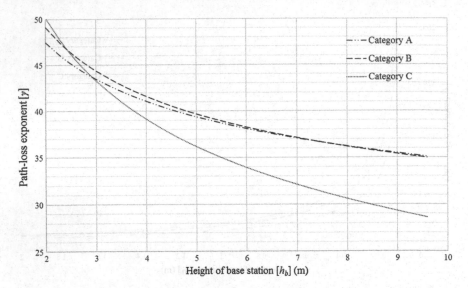

Fig. 5.8 Visual representation of the path-loss exponent of the basic SUI wireless transmission model with correction factors showing the variation in category attenuation (for Categories A, B and C) as a function of height of the base station (from 2 to a maximum of 10 m)

new method of calculating d_0) can also be utilized to determine propagation losses, albeit being a more complex alternative. This extended model is not reviewed in this chapter and is available in the reference provided at the beginning of this section.

The following model that is reviewed in this chapter is the COST-231 Hata semi-empirical model, applicable also for frequencies lower than the SUI Erceg model.

5.7.2   COST-231 Hata

Essentially a semi-empirical model and a amalgamation of the Walfisch-Bertoni and Ikegami path-loss models, the COST-231 Hata model is primarily recommended to be used for macro-cells (see Chaps. 1 and 4 of this book) in both urban and suburban environments.[2] Accuracy of this model is particularly good when the antenna of the transmitting source is located higher than the mean roof elevation in an urban or suburban area. In environments where the average roof height is similar, or higher than, the base station antenna, prediction errors increase significantly and poor performance could be experienced. The effective operating frequency range where this model is designed for ranges from approximately 800 MHz up to 2 GHz.

The COST-231 Hata model takes into account environmental variables such as the elevation of the adjacent buildings, width of the roads, the separation space between

---

[2]The COST-231 model is available online at https://www.xirio-online.com/help/en/cost231.html.

adjacent buildings, and the orientation of the road with respect to the direct path of the radio signal. These factors are similar to that of the SUI Erceg model with the primary difference being the implementation of these variables. The model can also distinguish between LoS and nLoS components and in the basic implementation, the total propagation loss is calculated as a sum of three components. These three components include

- free-space loss utilizing Friis free-space equations (L_{FS}),
- losses resulting from diffraction from adjacent rooftops and referred to the roof-to-street factor (L_{RTS}), and
- attenuation of the primary signal from multiple diffractions from rooftops along the path of the propagating signal (L_{MSD}).

The simplified model equation, based on the three listed components, can therefore be presented as

$$L_b[dB] = L_{FS}[dB] + L_{RTS}[dB] + L_{MSD}[dB] \qquad (5.15)$$

if $L_{RTS} + L_{MSD} > 0$ and

$$L_b[dB] = L_{FS}[dB] \qquad (5.16)$$

if $L_{RTS} + L_{MSD} \leq 0$. For the COST-231 Hata model, the free-space losses are calculated by

$$L_{FS}[dB] = 32.4 + 20\log d + 20\log f \qquad (5.17)$$

where d is the distance between the transmitting source and the receiving component specified in km and f is the operating frequency in MHz. As with the SUI model, the COST-231 model is limited to specific environments, and these boundary conditions include that the (in addition to the 800 MHz to 2 GHz operating frequency condition)

- antenna of the receiving component must have a height that ranges between 1 and 3 m,
- antenna of the transmitting source/base station must have a height that ranges between 4 and 50 m,
- space between the base station and the subscriber is between 20 cm and 50 km, and
- source must be higher than the receiver.

In contrast to the SUI model, the COST-231 Hata model is developed to produce a single model (as opposed a basic model, a modification employing correction factors, and an extended model). Albeit being more complex than the basic SUI model, the developed model incorporates all variables in its single form (assuming the environment is specified as either LoS or nLoS).

The LoS COST-231 Hata model is a low-complexity model than can easily be utilized and simulated. Notwithstanding the fact that very rarely perfect LoS is obtained in broadband wireless access communications, certain scenarios allow for LoS and therefore the model can be applied. Considering the development of last mile solutions in rural areas, hardware configurations can be deployed to allow for as much LoS communications as possible. In such cases, the path-loss model is defined by, comparable to the free-space losses in (5.17),

$$\overline{PL_{LoS}}(d)[dB] = 42.6 + 26 \log d + 20 \log f \qquad (5.18)$$

where f is again in MHz (<2000 MHz). A more complex nLoS COST-231 Hata version, and rightfully so, is described below. The nLoS COST-231 Hata model is derived from a typically scenario as shown in Fig. 5.9.

As shown in Fig. 5.9a and b, numerous variables and parameters contribute to signal propagation losses in an nLoS scenario when using the COST-231 Hata model. The combination of these parameters accounts for typical scenarios, similar to a street canyon, where the highest point of the buildings, base station, and subscriber antenna are used. The parameters defined in Fig. 5.9a and b are listed below:

- h_R is the mean height of the surrounding structures, specified in m,

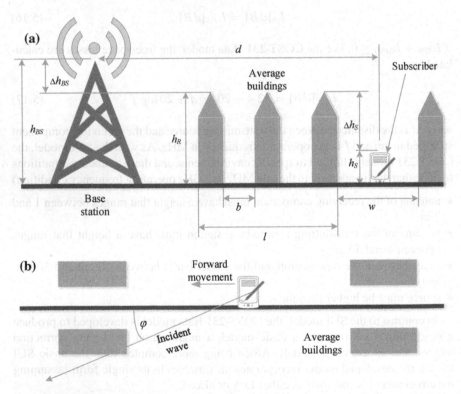

Fig. 5.9 A typical scenario describing the COST-231 Hata model in nLoS conditions representing **a** side view and **b** top view of the proposed environment

- h_{BS} is the height of the transmitting source (base station) antenna in m,
- w is the width of the street, also specified in m,
- b is the average separation among adjacent buildings in m,
- h_S is the height of the travelling subscriber antenna (approximately) in m,
- I is the total distance between the first and last buildings used in the scenario in m,
- d is the space between the source and the subscriber in km—assuming a static subscriber and therefore not accounting for Doppler shift,
- f is the operating frequency in MHz, and
- φ is the orientation angle between the direction of propagation and the street axis, specified in degrees.

Furthermore, with reference to Fig. 5.9a, the following equations hold true:

$$\Delta h_S(m) = h_r - h_S \tag{5.19}$$

and

$$\Delta h_{BS}(m) = h_{BS} - h_r. \tag{5.20}$$

The primary equations used to define the nLoS model for the COST-231 are given in (5.15) and (5.16), and individual components are defined such that the roof-to-street losses are

$$L_{RTS}[dB] = -8.2 - 10\log w + 10\log f + 20\log \Delta h_S + L_{ORI} \tag{5.21}$$

where L_{ORI} is a correction factor accounting for losses due to the physical orientation of the street. These correction factors are defined as

$$L_{ORI} = \begin{cases} -10 + 0.354\varphi, & 0° \le \varphi < 35° \\ 2.5 + 0.075(\varphi - 35), & 35° \le \varphi < 55° \\ 4.0 - 0.114(\varphi - 55), & 55° \le \varphi \le 90° \end{cases}. \tag{5.22}$$

Furthermore, the multipath diffraction losses are defined as

$$L_{MsD}[dB] = L_{BSH} + k_a + k_d \log d + k_f \log f - 9\log b \tag{5.23}$$

where L_{BSH} is an element that depends on the height of the transmitting source/base station and is given as

$$L_{BSH} = \begin{cases} -18\log(1 + \Delta h_{BS}), & h_b > h_r \\ 0, & h_b \le h_r \end{cases} \tag{5.24}$$

and the coefficients k_a (an rise in propagation attenuation for the source with antenna height below the average of the buildings), k_d (diffraction loss dependency on distance), and k_f (diffraction loss dependency on frequency) are given by

$$k_a = \begin{cases} 54, & h_b > h_r \\ 54 - 0.8\Delta h_b, & h_b \le h_r, d \ge 0.5 \\ 54 - 0.8\Delta h_b d/0.5, & h_b \le h_r, d < 0.5 \end{cases} \tag{5.25}$$

$$k_d = \begin{cases} 18, & h_{BS} > h_r \\ 18 - 15\frac{\Delta h_b}{h_r}, & h_{BS} \le h_r \end{cases} \tag{5.26}$$

$$k_f = \begin{cases} -4 + 0.7\left(\frac{f}{925} - 1\right), & * \\ -4 + 1.5\left(\frac{f}{925} - 1\right), & ** \end{cases} \tag{5.27}$$

where * represents a medium-sized city or residential area with medium foliage concentration, and ** represents a metropolitan urban area. In Fig. 5.10a the expected path-loss of the COST-231 Hata model for both LoS and nLoS communications. Both models represent an arbitrary 1.8 GHz broadband signal as a function of separation among the transmitting source and the receiving component. Figure 5.10b represents a constant arbitrary separation distance of 25 km with operating frequencies ranging from 800 MHz to 2 GHz, again for both LoS and nLoS communications.

In Fig. 5.10a the propagation losses for a 1.8 GHz broadband wireless signal as a function of separation among the transmitting source and the receiving component

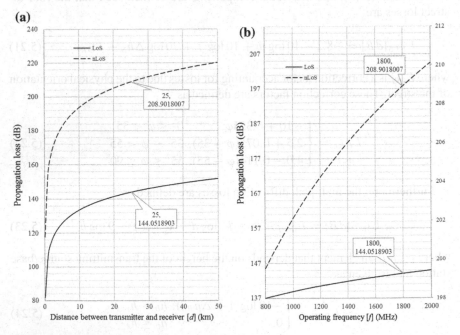

Fig. 5.10 Expected path-loss of the COST-231 Hata model for LoS and nLoS communications for a 1.8 GHz signal as a function of **a** separation among the transmitting source and the receiving component and **b** constant separation distance of 25 km with operating frequencies ranging from 800 to 2000 MHz (2 GHz) (nLoS on secondary axis)

is given. As expected, the nLoS model results in higher attenuation of the wireless signal (~208 dB at 25 km compared to ~144 dB). In Fig. 5.10b, the same model is applied for varying operating frequency, at a constant separation distance of 25 km. Again, the nLoS implementation yields a significantly larger path-loss across the entire frequency range.

The following model that is reviewed is the ECC-33 path-loss prediction model for signals in the low-GHz tier. The model is also related to the COST-231 Hata model in the following section.

5.7.3 ECC-33

The Electronic Communication Committee (ECC) established the ECC-33 path-loss prediction model that extrapolates from unique measured data by Okumura-Hata [16] and essentially adapted the results for a fixed wireless access system. The model is specifically adapted for medium- and large cities and includes "*correction factors for suburban and open areas*" [16]. The operating frequencies where this model is most effective range from approximately 700 MHz to 3.5 GHz. In this discussion, a comparative result (to the COST-231 Hata model) is plotted for operating frequencies between 800 MHz and 2 GHz. The ECC-33 path-loss model is simply

$$\overline{PL[dB]} = A_{FS} + A_{BM} - G_T - G_R \tag{5.28}$$

where A_{FS} is the free-space attenuation, A_{BM} is the elementary mean-path loss, G_T is the source elevation gain factor and G_R is the receiver antenna gain factor. Each of these factors are individually specified for the ECC-33 model such that

$$A_{FS} = 92.4 + 20 \log d + 20 \log f \tag{5.29}$$

where d and f are the distance between the source and the subscriber in km and the carrier frequency in GHz, respectively. The A_{BM} term is defined as

$$A_{BM} = 20.41 + 9.83 \log d + 7.89 \log f + 9.56 \left[\log f \right]^2 \tag{5.30}$$

and G_T is given by

$$G_T = \log \frac{h_{BS}}{200} \left[13.98 + 5.8 (\log d)^2 \right] \tag{5.31}$$

where h_{BS} is the height of the transmitting antenna. The ECC-33 defines G_R specifically for medium-sized urban areas as

$$G_R = \left[42.57 + 13.7 \log f \right] \left[\log h_S - 0.585 \right] \tag{5.32}$$

where h_S is the height of the receiver/subscriber antenna. For large cities, G_R is simplified and defined as

$$G_R = 0.759 h_S - 1.862. \qquad (5.33)$$

For a transmitting source/base station height of 15 m, a subscriber antenna elevation of 1 m, (5.28) is applied and the effects are offered in Fig. 5.11. The effects for a constant operating frequency (arbitrarily chosen as 1.8 GHz) at a varied separation distance as well as a constant separation distance (arbitrarily chosen as 25 km) with varied operating frequency are presented.

In Fig. 5.11a the propagation losses for a 1.8 GHz broadband wireless signal as a function of separation among the transmitting source and the receiving component is given. In Fig. 5.11b, the path-loss for a varied operating frequency at a constant 25 km separation distance is given. Additionally, the attenuation as estimated by the COST-231 LoS and nLoS models are also added to the figures. As noted, the ECC-33 model predicts the largest path-loss when compared to the COST-231 models, likely due to the correction factors for suburban environments. Comparing (5.29) with (5.17), there is a noticeable 60 dB attenuation added to the free-space attenuation term of the ECC-33 model. For a 1.8 GHz signal with 25 km between the transmitting source

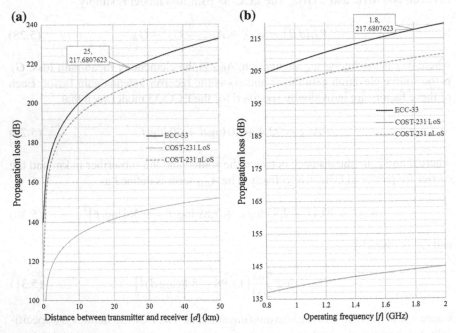

Fig. 5.11 Expected path-loss of the ECC-33 model for communications at **a** 1.8 GHz as a function of separation among the transmitting source and the receiving component and **b** constant separation distance of 25 km with operating frequencies ranging from 800 to 2000 MHz (2 GHz)

and the receiving component antenna, the ECC-33 model predicts an approximate 217 dB path loss, compared to the 144 and 209 dB for the COST-231 LoS and nLoS models, respectively.

In the following section, path-loss models for higher frequencies, moving into the mm-wave spectrum, are presented. These models are based on the lower tier models, but have additional loss-factors that account for high-frequency losses.

5.8 Fixed Wireless Access Path-Loss Models (mm-Wave Above 11 GHz)

The propagation loss models presented in the previous section, for frequencies typically below 11 GHz, form the fundamental basis of path-loss models at higher frequencies (above 11 GHz). However, with decreasing wavelength (increasing frequency), there are increased effects from the surrounding areas, as the wavelength becomes comparable to the size of particles in the environment. The additional attenuation experienced by high-frequency wireless signals, are complex to model individually and are typically measured and integrated into path-loss models as constant parameters. Parameters are fundamentally based on loss factors that influence primarily mm-wave signals, such as gaseous losses, precipitation, foliage, Doppler shift, free-space loss, scattering, and diffraction [8].

The first model reviewed is the blocked LoS (or nLoS) model, a simplified high-frequency model based on the FSPL model and capable of estimating attenuation in the mm-wave band.

5.8.1 Blocked LoS (nLoS)

The first mm-wave path-loss model reviewed is the simplified blocked LoS, an nLoS propagation model proposed in Chizhik et al. [3]. In this model, an additional 25 dB of attenuation is added to the free-space path-loss, accounting for shadowing loss from potential obstructions [4]. The path-loss model is therefore simply expressed as

$$\overline{PL}(d)[dB] = 20\log d + 20\log f + 20\log\left(\frac{4\pi}{0.3}\right) + 25 \qquad (5.34)$$

where f is the carrier in GHz. Noticeable from the blocked LoS path-loss model is its simplicity, since the additional potential losses are accounted for with a constant parameter which essentially estimates the worst-case attenuation across a specified distance. In Fig. 5.12, the attenuation characteristics of the blocked LoS model are

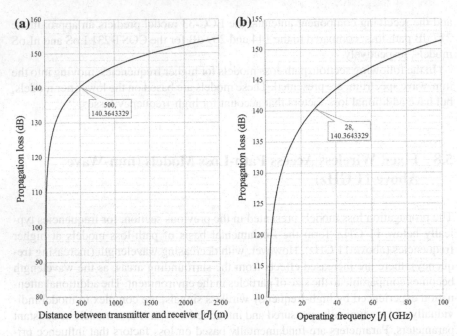

Fig. 5.12 Expected path-loss of the blocked nLoS model for communications at **a** 28 GHz as a function of separation among the transmitting source and the receiving component and **b** constant separation distance of 500 m with operating frequencies ranging from 0.5 to 100 GHz

presented for (a) a constant 28 GHz carrier frequency over a 2.5 km distance, and (b) a constant 500 m separation distance and a frequency span of 0.5–100 GHz.

In Fig. 5.12a the propagation losses for a 28 GHz broadband wireless signal as a function of separation distance are shown. In Fig. 5.12b, the path-loss for a varied operating frequency at a constant 500 m separation distance is given. For a constant 28 GHz carrier frequency, the attenuation based on the blocked LoS model ranges from approximately 85 to 155 dB across the 2.5 km span, whereas for a constant 500 m separation distance, across a 0.5–100 GHz span, the attenuation ranges between 105 and 150 dB.

The following path-loss models are based on a similar principle, albeit proposed for various scenarios and in certain circumstances based on the street canyon approximation. The first set of models is the 3GPP models, which includes the urban micro-cell (UMi)-street canyon approximations for LoS and nLoS as well as the urban macro-cell (UMa) approximations. The UMi scenarios are typically high-density open areas as well as street canyons where the base stations heights are below the rooftops. The subscriber antennas are near ground level, with heights around 1.5 m. UMa models are based on scenarios where the base station antennas are situated above rooftop level and subscriber antennas are again approximately near ground level. These models are also reviewed in further detail, and summarized, in Rappaport et al. [17], Niu et al. [15], and Haneda et al. [6].

5.8.2   3GPP

The 3rd Generation Partnership Project [1] is a collaborative effort among organizational partners, which include Nortel Networks in Canada, AT&T in North America, British Telecom in the United Kingdom, and France Telecom to create global system specification based on GSM networking protocols. The 3GPP standards have been developed for 2G, 2.5G, 3G, 4G and LTE, and currently for 5G networks and defines specifications from the user point of view, from support service requirements, and implementation of architectures by specifying applicable protocols. In the 5G spectrum, 3GPP attempts to provide models from 500 MHz (0.5 GHz) to 100 GHz based on modifications on the efforts to standardize models from 6 to 100 GHz. The first model proposed below is the UMi-street canyon LoS model. These models are subdivided into categories of separation between the transmitter and receiver, typically between 10 m and d'_{BP} and d'_{BP} and 5 km, where d'_{BP} is defined in Rappaport et al. [17] by

$$d'_{BP} = 4h'_{BS}h'_S \frac{f[GHz]}{c} \tag{5.35}$$

"where f is the carrier frequency in GHz, c is the speed of light ($\sim 300 \times 10^6$ m/s), and h'_{BS} and h'_S are the effective height of the base station and subscriber antenna, respectively", mathematically defined as

$$h'_{BS} = h_{BS} - 1m \tag{5.36}$$

$$h'_S = h_S - 1m \tag{5.37}$$

with h_{BS} and h_S being the actual base station and subscriber antenna heights, respectively. Furthermore, each model specifies a specific distance parameter, not necessarily the straight-line distance between the transmitting source and the receiving component. These parameters are defined in Fig. 5.13.

As shown in Fig. 5.13, the specifically defined distance parameters between the transmitting source and the receiving component are based on the height of each antenna and the separation distance. These parameters are used in the reviewed path-loss models, and Fig. 5.8 can be referenced to determine the distance being used. The inside-outside distance from the top of the source antenna to the uppermost point of the subscriber antenna is defined as

$$d_{3D-out} + d_{3D-in} = \sqrt{(d_{2D-out} + d_{2D-in})^2 + (h_{BS} + h_S)^2} \tag{5.38}$$

where all measurements are specified in meters. The first reviewed 3GPP model is the UMi-street canyon approximation for LoS, presented in the following subsection. Following this model are the UMi-street canyon for nLoS and the UMa LoS and nLoS models. Each model is presented in table-format for simplicity.

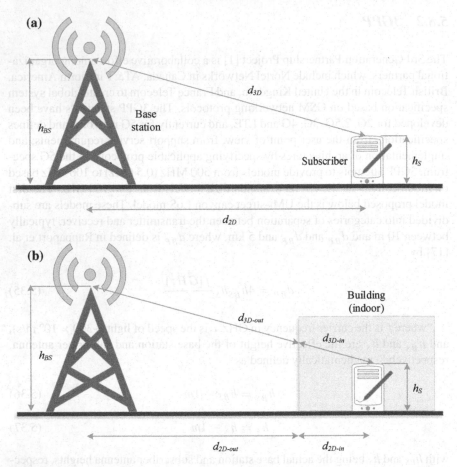

Fig. 5.13 Distance definitions for the 3GPP and 5GCM models presented in this section. **a** represents the definitions for d_{2D} and d_{3D} for outdoor models, whereas **b** defines $d_{2D\text{-}in}$, $d_{2D\text{-}out}$, $d_{3D\text{-}in}$, and $d_{3D\text{-}out}$ for indoor-outdoor models

Table 5.5 The 3GPP UMi-street canyon LoS model for $10 < d_{3D} < d'_{BP}$

Frequency range (f) [GHz]	Tx/Rx distance (d_{3D}) [m]	Subscriber antenna height (h_S) [m]	Base station height (h_{BS}) [m]	Shadowing (s) [dB]
$0.5 < f < 100$	$10 < d_{3D} < d'_{BP}$	$1.5 < h_S < 22.5$	10	4.00

5.8.2.1 UMi-Street Canyon (LoS)

See Table 5.5.

$$\overline{PL}(d)[dB] = 21 \log d_{3D} + 20 \log f + 32.4 \qquad (5.39)$$

Table 5.6 The 3GPP UMi-street canyon LoS model for $d'_{BP} < d_{3D} < 5000$

Frequency range (f) [GHz]	Tx/Rx distance (d_{3D}) [m]	Subscriber antenna height (h_S) [m]	Base station height (h_{BS}) [m]	Shadowing (s) [dB]
$0.5 < f < 100$	$d'_{BP} < d_{3D} < 5000$	$1.5 < h_S < 22.5$	10	4.00

Table 5.7 The 3GPP UMi-street canyon nLoS model for $d'_{BP} < d_{3D} < 5000$

Frequency range (f) [GHz]	Tx/Rx distance (d_{3D}) [m]	Subscriber antenna height (h_S) [m]	Base station height (h_{BS}) [m]	Shadowing (s) [dB]
$0.5 < f < 100$	$10 < d_{3D} < 5000$	$1.5 < h_S < 22.5$	10	8.20

Table 5.8 The 3GPP UMa LoS model for $10 < d_{3D} < d'_{BP}$

Frequency range (f) [GHz]	Tx/Rx distance (d_{3D}) [m]	Subscriber antenna height (h_S) [m]	Base station height (h_{BS}) [m]	Shadowing (s) [dB]
$0.5 < f < 100$	$10 < d_{3D} < d'_{BP}$	$1.5 < h_S < 22.5$	25	4.00

5.8.2.2 UMi-Street Canyon (LoS)

See Table 5.6.

$$\overline{PL}(d)[dB] = 40 \log d_{3D} + 20 \log f + 32.4 \qquad (5.40)$$

5.8.2.3 UMi-Street Canyon (nLoS)

See Table 5.7.

$$\overline{PL}(d)[dB] = 31.9 \log d_{3D} + 20 \log f + 32.4 \qquad (5.41)$$

5.8.2.4 UMa (LoS)

See Table 5.8.

$$\overline{PL}(d)[dB] = 22 \log d_{3D} + 20 \log f + 28.0 \qquad (5.42)$$

Table 5.9 The 3GPP UMa LoS model for $d'_{BP} < d_{3D} < 5000$

Frequency range (f) [GHz]	Tx/Rx distance (d_{3D}) [m]	Subscriber antenna height (h_S) [m]	Base station height (h_{BS}) [m]	Shadowing (s) [dB]
$0.5 < f < 100$	$d'_{BP} < d_{3D} <$ 5000	$1.5 < h_S < 22.5$	25	4.00

Table 5.10 The 3GPP UMa nLoS model for $10 < d_{3D} < 5000$

Frequency range (f) [GHz]	Tx/Rx distance (d_{3D}) [m]	Subscriber antenna height (h_S) [m]	Base station height (h_{BS}) [m]	Shadowing (s) [dB]
$0.5 < f < 100$	$10 < d_{3D} < 5000$	$1.5 < h_S < 22.5$	25	7.80

5.8.2.5 UMa (LoS)

See Table 5.9.

$$\overline{PL}(d)[dB] = 40 \log d_{3D} + 20 \log f + 28.0 \qquad (5.43)$$

5.8.2.6 UMa (nLoS)

See Table 5.10.

$$\overline{PL}(d)[dB] = 30 \log d_{3D} + 20 \log f + 32.4 \qquad (5.44)$$

5.8.3 Summary of the 3GPP Models

The 3GPP path-loss models are summarized in this section through plotting each model for a constant separation, either with varying carrier frequency, or with constant carrier frequency with varying separation distance. In Fig. 5.14, the results for varying carrier frequency with a 100 m separation distance are shown. The applicable models (UMi-street canyon and UMa models), are subdivided into categories based on the separation distance, therefore in Fig. 5.14, propagation loss for a distance of 100 m is based on (5.39), (5.41), (5.42), and (5.44).

As shown in Fig. 5.14, and expected from the path-loss models for the 3GPP models, specifically the UMi-street canyon approximation and the UMa LoS and nLoS models, for a separation distance of 100 m. The nLoS models present a noticeably higher attenuation compared to the LoS models, as can be expected. The UMi-street canyon approximation shows a somewhat higher attenuation when compared to its equivalent UMa models, for both the LoS and the nLoS models. In Fig. 5.15, a sim-

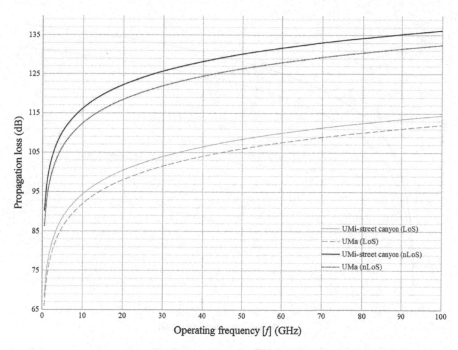

Fig. 5.14 Path-loss modeling of the 3GPP propagation models for frequencies between 0.5 and 100 GHz for a fixed antenna separation distance of 100 m

ilar comparison is made between the UMi-street canyon LoS and UMa LoS models, when the separation distance falls outside d'_{BP}, arbitrarily chosen as 2500 m (2.5 km) and therefore applicable to (5.40) and (5.43).

As shown in Fig. 5.15, and expected from the path-loss models for the 3GPP models, the UMi-street canyon LoS model is again subject to a higher attenuation estimation compared to the UMa LoS model. Propagation loss for the UMi-street canyon model ranges from approximately 163 dB at 500 MHz up to approximately 208 dB at 100 GHz.

In Figs. 5.16 and 5.17, the 3GPP models are presented for a constant 28 GHz carrier frequency, for separation spaces ranging from 10 up to 5000 m. Since the UMi-street canyon and UMa models are subdivided based on separation distance, Fig. 5.16 shows the models' relevance for distances between 10 and 100 m (d'_{BP} in this scenario), whereas Fig. 5.17 shows the models' relevance for distances between 100 and 5000 m (5 km).

As shown in Fig. 5.16, for relatively short distances between the transmitter and the receiver (100 m), UMi-street canyon nLoS presents the highest attenuation estimation of between 93 and 125 dB from 10 to 100 m at a constant 28 GHz carrier frequency. The UMa LoS presents the lowest attenuation estimation ranging from 82 to 103 dB. At $d'_{BP} = 100$ m, a discontinuity occurs, and the propagation models are presented in Fig. 5.17 for these ranges.

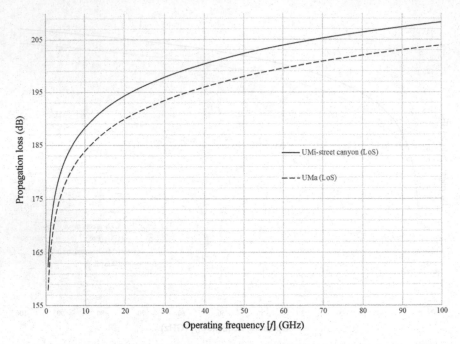

Fig. 5.15 Path-loss modeling of the 3GPP propagation models for frequencies between 0.5 and 100 GHz for a fixed antenna separation distance of 2500 m (2.5 km)

In Fig. 5.17, the attenuation estimation characteristics for the 3GPP models are presented for a constant 28 GHz carrier frequency, for a range of 100 m (d'_{BP}) to 5000 m. Noticeable from Fig. 5.17 is that the UMi-street canyon and UMa LoS models present the largest attenuation estimation over the 100–5000 m range. Although this appears to be incorrect, as one would expect the nLoS models to present the highest attenuation estimations, it has been verified in the June 2018 update [1] and assumed to be correct.

5.8.4 5GCM

The 5G mm-wave channel models (5GCM) initiated by the United States National Institute of Standards and Technology (NIST) are based on the 3GPP-3D channel model and implement multi-frequency channel practical measurement and ray-trace modeling methods. Measured data (which are still in progress) for path-loss, dispersion losses, and obstruction models for nLoS scenarios were obtained and the models proposed. The frequency spectra where channel measurement are obtained and models proposed primarily include sub-6, 10–11, 14–15, 18, 26, 28–29, 38–40, 45 GHz, E-band (71–76 and 81–86 GHz), and the 60 GHz band.

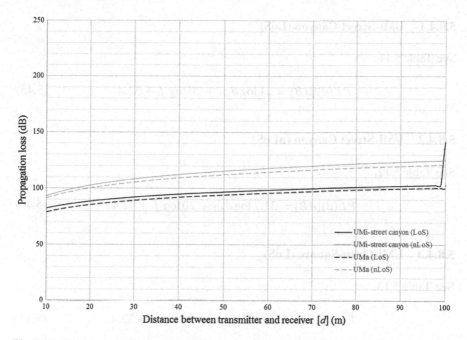

Fig. 5.16 Path-loss modelling of the 3GPP propagation models for separation distances between 10 and 100 m and a constant 28 GHz carrier frequency

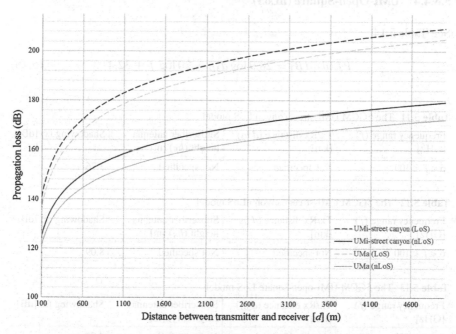

Fig. 5.17 Path-loss modeling of the 5GCP propagation models for separation distances between 1 and 5000 m (5 km) and a constant 28 GHz carrier frequency

5.8.4.1   UMi-Street Canyon (LoS)

See Table 5.11.

$$\overline{PL}(d)[dB] = 21 \log d_{3D} + 20 \log f + 32.4 \qquad (5.45)$$

5.8.4.2   UMi-Street Canyon (nLoS)

See Table 5.12.

$$\overline{PL}(d)[dB] = 31.7 \log d_{3D} + 20 \log f + 32.4 \qquad (5.46)$$

5.8.4.3   UMi-Open-Square (LoS)

See Table 5.13.

$$\overline{PL}(d)[dB] = 18.5 \log d_{3D} + 20 \log f + 32.4 \qquad (5.47)$$

5.8.4.4   UMi-Open-Square (nLoS)

See Table 5.14.

$$\overline{PL}(d)[dB] = 28.9 \log d_{3D} + 20 \log f + 32.4 \qquad (5.48)$$

Table 5.11   The 5GCM UMi-street canyon LoS model

Frequency range (f) [GHz]	Tx/Rx distance (d_{3D}) [km]	Subscriber antenna height (h_S) [m]	Shadowing (s) [dB]
$6 < f < 100$	Not specified	Not specified	3.76

Table 5.12   The 5GCM UMi-street canyon nLoS model

Frequency range (f) [GHz]	Tx/Rx distance (d_{3D}) [km]	Subscriber antenna height (h_S) [m]	Shadowing (s) [dB]
$6 < f < 100$	Not specified	Not specified	8.09

Table 5.13   The 5GCM UMi-open-square LoS model

Frequency range (f) [GHz]	Tx/Rx distance (d_{3D}) [km]	Subscriber antenna height (h_S) [m]	Shadowing (s) [dB]
$6 < f < 100$	Not specified	Not specified	4.20

Table 5.14 The 5GCM UMi-open-square nLoS model

Frequency range (f) [GHz]	Tx/Rx distance (d_{3D}) [km]	Subscriber antenna height (h_S) [m]	Shadowing (s) [dB]
$6 < f < 100$	Not specified	Not specified	7.10

Table 5.15 The 5GCM UMa LoS model

Frequency range (f) [GHz]	Tx/Rx distance (d_{3D}) [km]	Subscriber antenna height (h_S) [m]	Shadowing (s) [dB]
$6 < f < 100$	Not specified	Not specified	4.10

Table 5.16 The 5GCM UMa nLoS model

Frequency range (f) [GHz]	Tx/Rx distance (d_{3D}) [km]	Subscriber antenna height (h_S) [m]	Shadowing (s) [dB]
$6 < f < 100$	Not specified	Not specified	6.80

5.8.4.5 UMa (LoS)

See Table 5.15.

$$\overline{PL}(d)[dB] = 20 \log d_{3D} + 20 \log f + 32.4 \tag{5.49}$$

5.8.4.6 UMa (nLoS)

See Table 5.16.

$$\overline{PL}(d)[dB] = 30 \log d_{3D} + 20 \log f + 32.4 \tag{5.50}$$

5.8.5 Summary of the 5GCP Models

The 5GCP path-loss models are summarized in this section through plotting each model for a constant separation distance between the transmitting source and the receiving component, either with varying carrier frequency, or with constant carrier frequency with varying separation distance. In Fig. 5.18, the results for varying carrier frequency with a 100 m separation distance is shown.

As shown in Fig. 5.18, and expected from the path-loss models for the 5GCP models, the primary difference in the propagation losses is based on either the constant factor added to the path-loss model, or the scenario, whether it is LoS or nLoS. The LoS models all present a significantly smaller attenuation of between approximately 90 dB at 6 GHz (maximum for all the models) up to approximately 115 dB at

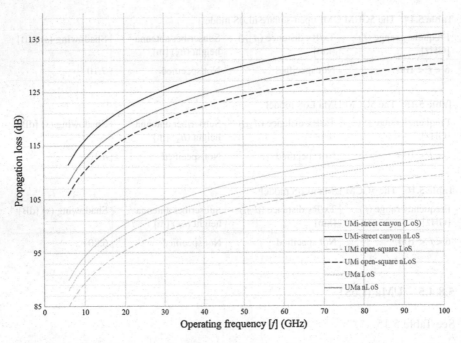

Fig. 5.18 Path-loss modeling of the 5GCP propagation models for frequencies between 6 and 100 GHz for a fixed antenna separation distance of 100 m

100 GHz. The nLoS ranges from approximately 111 dB at 6 GHz (UMi-street canyon model) upwards to approximately 135 dB at 100 GHz. In Fig. 5.19, the same models are presented at a carrier frequency of 28 GHz, with varying separation between 1 m (logarithmic results are inaccurate for distances shorter than 1 m) and 5 km.

A similar result is obtained in Fig. 5.19, with a constant 28 GHz carrier frequency and a varied separation distance between 0 m and 5 km. The nLoS experiences the higher attenuation, up to approximately 180 dB for the UMi-street canyon approximation at a separation distance of 5 km, whereas the LoS model has a maximum attenuation of approximately 140 dB at 5000 m.

5.9   Additional mm-Wave Design Considerations

Outlined in Hemadeh et al. [8], several additional techniques can be employed to construct an mm-wave communication channel that is capable of transmitting and receiving information effectively and efficiently, despite its primary caveat, the high free-space losses because of its short wavelength. The characteristics defined in Hemadeh et al. [8] include

● using at least one antenna array capable of beamforming,

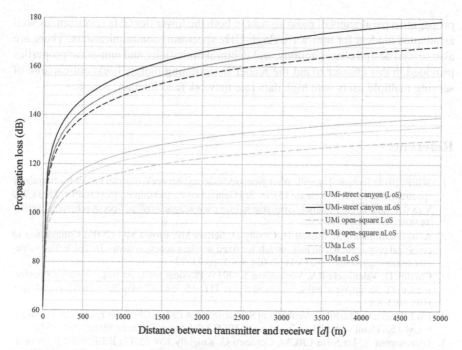

Fig. 5.19 Path-loss modeling of the 5GCP propagation models for separation distances between 1 m and 5 km and a constant 28 GHz carrier frequency

- enhancing spectral efficiency by stacking numerous antenna elements in a MIMO configuration,
- limiting attenuation and increasing capacity by reducing the size of the cells,
- taking advantage of the potential high sampling frequencies and frequency-selective channels (fading) in mm-wave signals,
- using advanced and efficient modulation and duplexing techniques, and
- sharing second-order spatial characteristics to counteract sparse multipath scattering (multi-user grouping).

Reviewing each of these techniques falls outside the scope of this chapter, and Hemadeh et al. [8] provide sufficient literature on each technique.

5.10   Conclusion

This chapter presents a theoretical analysis terrestrial and mm-wave mobile backhaul as a proposed solution for last mile internet access. The focus of this chapter is to present the characteristics of mm-wave electromagnetic signals, as well as the limitations and challenges of using the technology to deploy mobile networks in rural or urban areas. This is primarily achieved by presenting path-loss models and propagation parameters of radio signals in both the lower tier GHz spectrum, as well as adapted models for mm-wave (high-GHz spectrum) communications. There are also several design considerations, architectural limitations, and mm-wave properties provided in this chapter to aid the reader in designing mobile backhaul capable of serving multiple users with high data rate network (and internet) access.

References

1. 3GPP (2018) Technical specification group radio access network; channel model for frequency spectrum above 6 GHz (Release 15). In: 3rd Generation Partnership Project (3GPP), TR 38.900 V15.0.0, June 2018. [Online]. Available at http://www.3gpp.org/ftp//Specs/archive/38_series/38.900/
2. Abhayawardhana VS, Wassell IJ, Crosby D, Sellars MP, Brown MG (2005) Comparison of empirical propagation path loss models for fixed wireless access systems. In: IEEE 61st vehicular technology conference (VTC 2005), vol 1, pp 73–77
3. Chizhik D, Valenzuela RA, Venkatesan S (2015) Physical limits on beam switching performance of LOS mmWave links. Technical report. ITD-15-55823C, May 2015. Bell Labs, Murray Hill, NJ, USA
4. Du J, Onaran E, Chizhik D, Venkatesan S, Valenzuela A (2017) Gbps user rates using mmWave relayed backhaul with high-gain antennas. IEEE J Sel Areas Commun 35(6):1363–1372
5. Ghasempour Y, Da Silva CRCM, Cordeiro C, Knightly EW (2017) IEEE 802.11 ay: next-generation 60 GHz communication for 100 Gb/s Wi-Fi. IEEE Commun Mag 186–192
6. Haneda K, Tian L, Zheng Y, Asplund H, Li J, Wang Y, Steer D, Li C, Balercia T, Lee S, Kim Y, Ghosh A, Thomas T, Nakamura T, Kakishima Y, Imai T, Papadopoulas H, Rappaport TS, MacCartney GR, Samimi MK, Sun S, Koymen O, Hur S, Park J, Zhang C, Mellios E, Milosch AF, Ghassamzadeh SS, Ghosh A (2016) 5G 3GPP-like channel models for outdoor urban microcellular and macrocellular environments. In: IEEE 83rd vehicular technology conference (VTC 2016), May 2016
7. Hassibi B, Hochwald BM (2003) How much training is needed in multiple-antenna wireless links? IEEE Trans Inf Theory 49:951–963
8. Hemadeh IA, Satyanarayana K, El-Hajjar M, Hanzo L (2018) Millimeter-wave communications: physical channel models, design considerations, antenna constructions, and link budget. IEEE Commun Surv Tutor 20(2):870–913
9. Jung H, Lee I (2016) Outage analysis of millimeter-wave backhaul in the presence of blockage. IEEE Commun Lett 20(11):2268–2271
10. Kim S, Visotsky E, Moorut P, Bechta K, Ghosh A, Dietrich C (2017) Coexistence of 5G with the incumbents in the 28 and 70 GHz bands. IEEE J Sel Areas Commun 35(6):1254–1268
11. Lambrechts JW, Sinha S (2016) Microsensing networks for sustainable cities. Springer International Publishing. ISBN 978-3-319-28357-9
12. Lambrechts JW, Sinha S (2017). SiGe-based re-engineering of electronic warfare subsystems. Springer International Publishing. ISBN 978-3-319-47402-1
13. Li X, Yu J, Xiao J (2016) Demonstration of ultra-capacity wireless signal at W-band. J Lightwave Technol 34(1):180–187
14. Nie S, MacCartney GR, Sun S, Rappaport TS (2014) 28 GHz and 73 GHz signal outage study for millimeter wave cellular and backhaul communications. In: IEEE ICC 2014—wireless communications symposium, pp 4856–4861
15. Niu Y, Li Y, Jin D, Su L, Vasilakos AK (2015) A survey of millimeter wave communications (mmWave) for 5G: opportunities and challenges. Wirel Netw 21(8):2657–2676
16. Rappaport TS (2002) Wireless communications: principles and practice, 2nd edn. Prentice Hall. ISBN 0-471-65596-1

17. Rappaport TS, Xing Y, MacCartney GR, Molisch AF, Mellios E, Zhang J (2017) Overview of millimeter wave communications for fifth-generation (5G) wireless networks-with a focus on propagation models. IEEE Trans Antennas Propag. Special Issue on 5G. arXiv:1708.02557
18. Rasekh ME, Farzaneh F, Shishegar AA (2010) A street canyon approximation model for the 60 GHz propagation channel in an urban environment with rough surfaces. In: 5th international symposium on telecommunications (IST 2010), pp 132–137
19. Sharma PK, Singh RK (2010) Comparative analysis of propagation path loss models with field measured data. Int J Eng Sci Technol 2(3):2008–2013
20. Sulyman AI, Alwarafy A, MacCartney GR, Rappaport TS, Alsanie A (2016) Directional radio propagation path loss models for millimeter-wave wireless networks in the 28-, 60-, and 73-GHz bands. IEEE Trans Wirel Commun 15(10):6936–6947
21. Sun S, Rappaport TS, Rangan S, Thomas TA, Ghosh A, Kovacs IZ, Rodriguez I, Koymen O, Partyka A, Jarvelainen J (2016) Propagation path loss models for 5G urban micro- and macro-cellular scenarios. In: IEEE 83rd vehicular technology conference (VTC2016), pp 1–6
22. Vardoulakis S, Fisher BEA, Pericleous K, Gonzalez-Flesca N (2003) Modelling air quality in street canyons: a review. Atmos Environ 37:155–182
23. Zhang G, Quek TQS, Kountouris M, Huang A, Shan H (2016) Fundamentals of heterogeneous backhaul design—analysis and optimization. IEEE Trans Commun 64(2):876–889

Chapter 6
Successful Implementations of Last Mile Internet Solutions in Emerging Markets

Abstract Emerging markets have in many instances realized the importance of last mile solutions and applied it to boost socioeconomic growth and create solutions that embrace the fourth industrial revolution. The internet is an enabler of sustainability, even in regions that are geographically distant from industrial hubs such as large cities. Acknowledging these initiatives, identifying the challenges faced and how these challenges were overcome, as well as recognizing the technologies used to implement these solutions is important for companies, especially for local small and medium-sized enterprises, to learn from and improve new innovative last mile solutions to connect the last billion. Training the local populations and increasing the number of skilled workers in emerging markets, together with researching and analyzing successful last mile implementations provides the tools to eradicate the digital divide, empower local people, spearhead sustainable socioeconomic growth and prepare the emerging world to participate in the fourth industrial revolution. This chapter researches successful last mile implementations in emerging markets and presents the successes and failures of each of these solutions.

6.1 Introduction

The digital divide, and the lack of high quality last mile solutions in emerging markets and rural areas is preventing societies from taking advantage of the benefits of information sharing [16]. To create inclusive and sustainable connectivity, physical infrastructure and long-term management of skills are required, with strong leadership on global and local levels crucial to its success. A lack of digital infrastructure in emerging economies leads to a lack in network connection, devices, software, and applications [16]. Accompanied with a lack of digital skills that create or add value to services and digital products, an inclusive digital economy is not possible. According to Shengling et al. [16], the full benefits of information and communication technology (ICT) are, in summary,

- capable of promoting efficiency through automation and access to information and knowledge,

© Springer Nature Switzerland AG 2019
W. Lambrechts and S. Sinha, *Last Mile Internet Access for Emerging Economies*,
Lecture Notes in Networks and Systems 77,
https://doi.org/10.1007/978-3-030-20957-5_6

- a promoter of socioeconomic inclusiveness through information sharing and transparency, and
- an enabler of new economies in line with the fourth industrial revolution.

Furthermore, Shengling et al. [16] proposes a framework for analyzing supply-side ICT policies that supply internet connectivity to a country (the first mile), as it is distributed through the country (the middle mile), where and how it reaches the consumer (the last mile), and mitigating ad hoc issues in a country (known as the invisible mile). Through this framework, Shengling et al. [16] aims to

- eliminate capitalist monopolies that monetize the first mile and liberalizing the ICT market for decentralized connectivity,
- encouraging open access for backhaul networks in the middle mile,
- encouraging last mile access from governmental policies and making access lines available to competitors at wholesale prices,
- in terms of the invisible mile, ensuring competitive access to spectrum through inclusive and fair policies, and
- most importantly, having affordable alternatives ranging from the first mile to the last mile, and in between, essentially addressed through free and fair competitive policies.

Furthermore, the framework in Shengling et al. [16] includes addressing digital investments in industry innovations in the ICT sector to ensure infrastructure sustainability. In addition, Shengling et al. [16] proposes implementing dynamic education to teach skills that are defined by Industry 4.0 (even if the skills requirements do not yet exist) and arming workers with knowledge of the ICT sector that are required to make decisions using the technology.

Emerging markets are also encouraged to be an integral part of the fourth industrial revolution, as growth in this sector (ICT specifically), is fast compared to most other sectors. Globally, the revenue of the ICT market is estimated at over EUR 4.4 trillion by the end of 2019 [17], and Fig. 6.1 presents the growth of this sector since 2005.

As seen from Fig. 6.1, there has been a steady growth in the ICT market from 2005 (valued at EUR 2.3 trillion) each year up to the recorded data in 2016. The estimated market value for the end of 2019 is at EUR 4.46 trillion, a 12.2% increase from 2016. The average per annum growth over this period is 5.69% and the data includes that of the entire ICT market, television, and video services. In South Africa, an emerging market, and part of the BRICS nations, the telecommunications industry has grown from approximately R137 billion (US $10 billion) in 2006, to almost R315 billion (US $23.5 billion) in 2016. These figures were reported at the end of 2018 by Stats SA [18], with a steady per annum growth of 5.9% over this period—in line with the global reports [17]. In terms of employment, the telecommunications industry in South Africa employed 58,407 people in 2016, up from 50,722 in 2006, although the highest number of persons employed in the telecommunications industry was reported in 2013, at 66,408 people [18]. A statistic of significance in Stats SA [18] is the 2016 income versus employment of the telecommunications industry and the post and associated courier service (could be considered the historical telecommunications sector). The reported statistics are summarized in Table 6.1.

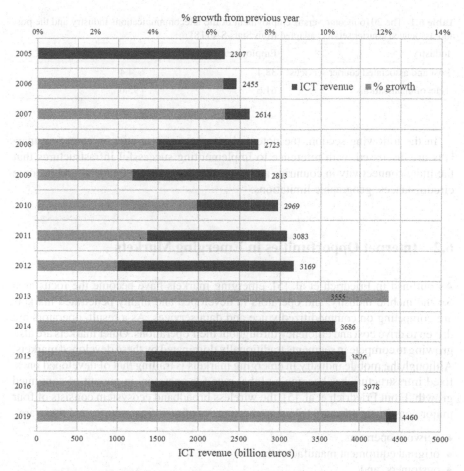

% growth from previous year

Fig. 6.1 The ICT sector revenue, in billion euros, and per annum growth in % from 2005 and estimated at the end of 2019, adapted from Statista [17]

As shown in Table 6.1, when comparing the post and associated courier services and telecommunications industries, the employment of the post and associated courier industries still has a significant contribution, at 38.4%, with telecommunications employment at 61.6% (if only comparing these two industries with each other). However, the income contribution of these industries differs vastly, with post and associated courier facilities only contributing 4.4% of total income, and telecommunications contributing the majority at 95.6%. This shows that there is a significant drive for youth and skilled workers to enter the telecommunications market, and ICT in general, as most of the opportunities in terms of financial stability lies in this sector (when compared to the traditional post and associated courier facilities).

Table 6.1 The 2016 income versus employment of the telecommunications industry and the post and associated courier service, adapted from Stats SA [18]

Industry	Employment (%)	Income (%)
Post and associated courier services	38.4	4.4
Telecommunications	61.6	95.6

In the following section, the potential internet opportunities in emerging markets are discussed, with reference to implementing successful infrastructures that facilitate connectivity in countries and areas that have severe financial and in some circumstances, geographic limitations.

6.2 Internet Opportunities in Emerging Markets

As outlined in Friedrich et al. [5], emerging markets have become the focus area for the mobile industry, and operators in developed and highly penetrated markets are competing on commoditized voice and data services. As a result, operators are driven to drive cost and efficiency throughout their operations. Other markets are also growing to compete in segments traditionally dominated by these developed markets. Although the mobile industry in emerging markets is trailing that of developed ones, fixed infrastructure has become a global race for operators in pursuit of continued growth. From Friedrich et al. [5], the wireless broadband ecosystem consists of four major constituents that contribute to the success of an infrastructure, and these are

- network operators,
- original equipment manufacturers,
- customers, and
- device manufacturers.

In 2007, it was assumed that clients in developing countries would first look to fulfil the most elementary communication essentials, being voice. In an evolving mobile industry, voice has become less prevalent, although still the top priority, access to affordable data has become equally important. Although internet usage in emerging markets is still relatively low, the internet ecosystem in these countries are underdeveloped with few local servers and content offerings tailored to local needs.

Wireless services have an inherent advantage over fixed-line networks and is arguably more economical for offering mid-band (~10 Mbps) broadband access. Recent advances in technology such as fiber and mm-wave backhaul have changed the broadband environment significantly. These have become heavily backed by major vendors and standardization institutions. Their success however relies not only on the technology performance, but also on the underlying ecosystems that are in place (which are often underdeveloped). While there is a solid social dispute for getting more people in emerging markets online, there is a similarly compelling economic argument. Research has revealed that for each added 10% points of internet infiltration, 1.2% points is added to the per capita GDP development in developing countries, and each added 10% points of broadband infiltration adds 1.38% points of per capita GDP development [13]. According to Becker [1], the emerging markets that embrace emerging technologies, can potentially outperform developed markets, a trend that has been witnessed [1] and is likely to continue. From Becker [1], global investment in emerging and frontier economies has been a major theme over the past two decades, primarily from the Chinese economy pushing capital into these markets during the 1990s. During the 2000s, the flourishing US and Europe housing markets helped stimulate a global commodity as well as energy investment uptake. Together, these factors secured fast growth in emerging economies, leading to a growth in middle-class citizens and an overall lowering in political risk [1]. It should however be noted that from around 2012, global investment into emerging markets and the associated socio-economic growth, was stifled by worldwide decline in markets, an unfortunate series of events as Industry 4.0 was set to take off. Last mile connectivity and capital investment into providing internet to emerging and rural economies also suffered from this slowdown, and sources of financial backing declined, leaving emerging markets to obtain and sustain their own investments. With emerging economies expected to account for almost 70% of global GDP in 2030 [3], opportunities for consumers in these markets remain strong, and many services and products will be associated with Industry 4.0 and inherent last mile solutions over this time. For investors spending in emerging markets, there are unfortunately risks accompanying these endeavors, and Boumphrey and Verikaite [3] presents guidelines, or risk factors, when working with developing markets. According to Boumphrey and Verikaite [3], identifying these factors and mitigating any risks that may arise are important to ensure sustainable investments, and these factors are adapted and summarized below:

- Governments and industries in financially restricted economies typically employ outdated methodologies, a lack of technological solutions, and in many instances, corrupt methods of gathering and supplying statistical data of their financial and economic performance. If for example, distribution of last mile access in rural areas were undertaken, working with a market research provider to determine high-quality data sets would be beneficial as opposed to using locally supplied data.
- Related to the quality of data in emerging economies, data availability is also generally poor, or not available. As a result, proxies can assist in gaining a better understanding and context of a specific market. For example, there might be a lack of data on the percentage of households in a given area with access to broadband internet. It is however possible, with reliable estimates, to construct good quality and widely available data on broadband internet users based on local internet service providers' (ISPs') annual financial and user reports.
- In all markets, data on technology trends are changing rapidly, and can vary significantly over a six-month period. In emerging markets, surveys could be carried out sporadically and over extended period. As a result, wariness of data capture

timelines is important, especially when linked to technology-enabling services such as last mile internet access.

- Not only should data be recent, but also it should be region-specific. It is a common characteristic of emerging markets that there is a large wage gap between the poor and the wealthy, and each group typically concentrate in specific areas. Therefore, what is true of internet/technology access in an area could be vastly different to another, even if the proximity between these areas are relatively small.
- Boumphrey and Verikaite [3] furthermore highlights factors such as clear distinguishing of definitions in emerging markets, which could also be vastly different to that of developing countries. As an example, access to the internet in a developed country might inadvertently mean access to a smart phone, whereas in a rural area, this might refer to having access to a local internet café, several kilometers from the individual's dwelling, and at overly expensive rates. Surveys and market research should therefore have very specific terminology and it should not be assumed that these refer to the same services or products everywhere in the world.

Finally, Boumphrey and Verikaite [3] propose that all data sourced in emerging markets, without prejudice, must be questioned and assessed for quality, political bias, clarity of definitions, as well as the accuracy of the methodology implemented to gather the data. These suggestions might seem overly cautious, but it is important to highlight the sheer volume of people not connected to the internet, and the primary reasons for it, as discussed in the following section.

6.3   The Last Billion

With over half of the world's population not connected to the internet, innovative and low-cost business models are critical in enabling the world's bottommost paid customers to contribute in the incipient digital economy, Industry 4.0. This lowest-income population is often referred to as the last billion, and have an average income of US \$45, and a potential affordable communications expenditure of US \$2.25 per months. The majority of this population group lives in Sub-Saharan Africa and on the subcontinent of India. Roughly 75% residing in rural regions that are predominantly dependent on agriculture as a source of livelihood. An estimated 20–40% live without even the most basic mobile communications [14]. In Schmida et al. [14], the monthly communications expenditure for the world's populace is tabulated, and presented in Table 6.2. Segmenting the world's isolated (in terms of internet connectivity) by income level and bottommost income population sets a strong target that business models must address as to approach viability for this group.

From Table 6.2, the market opportunity for the communication needs of the last billion is estimated at around US \$27 billion, a significant opportunity for developed and penetrated markets to engage with emerging markets to bring last mile solutions to this population group. Again, low-cost and tailored solutions are required to meet the needs of this group, as the affordable monthly communications spend is low

(US $2.25). Furthermore, if business models can be successful with this group, it is logical that they have a worthy likelihood of flourishing at the upper price points (5th and 6th billion population groups). Highlighted by Schmida et al. [14] is the fact that consumers generally do not care about the technology behind how they access the internet, as long as the business model is tailored in a way that is affordable to the consumer. Last mile connectivity is therefore a strong function of the economic drive of internet connectivity, and in this book, the available technologies are presented to achieve this. Whether an end-user device is using 4G, Wi-Fi, or TVWS is typically irrelevant to the consumer, also if whether the backhaul is provided by fiber, mm-wave, of satellite, does not matter to the consumer in general. From this perspective, Schmida et al. [14] presents several key findings, further strengthening the findings of [3], which are adapted and listed below:

- business models and current approaches to providing internet to users must be reimagined for the last billion of the global population,
- the technologies to achieve this exist, the challenges lie in deploying and scaling business models that are capable to sustainable provide internet to low-income consumers,
- the innovative and scaled business models also exist, and network operators need to adapt to emerging markets' needs and buying power,
- revenue streams must be diversified, costs lowered, and local entrepreneurs should serve as advocates to drive income through value-added services,
- partnership opportunities and adapted regulations are required for these models to succeed.

Schmida et al. [14] further reports that the presented study follows three steps that seize the responses from evaluated partakers, and to use these inputs as a baseline to analyze the opportunities, challenges, and potential ways forward to connect the last billion. The three steps followed in Schmida et al. [14] are to

- determine and present certain major developments as well as trends that are shaping the modern market environment, especially in terms of connectivity ventures,

Table 6.2 The monthly communications expenditure for the world's populace, adapted from Schmida et al. [14]

Population group	Average annual income (US $)	Affordable monthly communications spend (US $)	Total attainable market (US $ billion)
1st billion	29,206	205	2460
2nd billion	12,722	53	636
3rd billion	5540	23	276
4th billion	2987	12	144
5th billion	1771	7	84
6th billion	1065	4.40	53
7th (last) billion	540	2.25	27

- present distinctive emerging solutions which enable last billion business models illustrated through five case studies, and
- review the challenges and recommendations to scale innovative connectivity business models in emerging markets.

Regarding the first step presented in Schmida et al. [14], specific reference is made to the challenges and opportunities that internet service providers (ISPs) and mobile network operators (MNOs) are faced with in terms of their willingness to reach consumers in remote (rural) areas. There is a clear indication of the consensus among major corporations that ubiquitous internet access should be addressed, albeit the primary and most difficult challenge to overcome. To their advantage, the growing consumption of mobile services by all income echelons, as their principal device to access the internet, redefines this market and allows ISPs and MNOs to take advantage of the increasingly ubiquitous methods of accessing their services. Furthermore, Schmida et al. [14] reports that it is possible to build low-cost networks using existing technology and provide connectivity for a large percentage of consumers, and mature technologies such as 3G and Wi-Fi could be deployed more effectively. It is possible to deliver high quality connectivity experience to users by combining these technologies with low-cost base transceiver stations (BTS) and emerging technologies such as TVWS. Additionally, to improve on the quality of service (QoS), service providers can minimize bandwidth usage by keeping the content local, implementing efficient traffic routing strategies, and optimizing upload and download times. Renewable energy such as solar power and wind energy is also a critical investment to ensure consistent energy supply.

A clear reference to investing in network infrastructure, specifically backhaul, is presented in the second step proposed by Schmida et al. [14] with respect to emerging solutions. Investment in backhaul has proportionate influence on the feasibility of last mile business models [14]. The report refers to airborne network infrastructures to provide connectivity directly to consumers via aerial devices such as satellites or balloons. However, Schmida et al. [14] also found that this strategy might be the least viable solution, since there will still be a need of a ground-based infrastructure and its cost-effectiveness and relevancy are difficult to determine. It is therefore also referenced that fiber optic last mile solutions are possibly the most feasible mainstay backhaul solution, capable of routing large volumes of traffic and having very little susceptibility to interference.

Finally, Schmida et al. [14] presents the challenges experienced in emerging markets that stifle last mile connectivity, highlighting issues such as

- scarce initial funding and limited investments from government and local investors,
- inaccessible networks and restricted (and often absent) information sharing between potentially collaborative partners, and
- very little backing from regulatory bodies and policy makers.

It is evident from these findings by Schmida et al. [14], that the issues plaguing emerging markets to roll out last mile connectivity, coincides with the findings not only of this book, but also that of Boumphrey and Verikaite [3]. Schmida et al. [14] do present certain tools for scaling the last billion business models, which includes suggestions to pre-screen viable investments, early stage concessionary financing, near commercial financing, de-risking instruments, and ample knowledge sharing and technical assistance between successful projects and new endeavors. Each project and investment strategy and policies are different, and this chapter highlights certain successful last mile investments in emerging markets to assist in determining how these successful policies and strategies were structured. Additionally, the challenges and limitations reviewed in this chapter serve as identified risks that should be mitigated if identified in a certain area. The following sections reviews several industry innovations in last mile connectivity in emerging markets.

6.4 Industry Innovations in Last Mile Connectivity

Several industry innovations from large and small corporations aims to prototype and test solutions to provide internet connectivity to areas where traditional implementations are not feasible, or not possible at all. This section reviews implementations from companies such as

- Google,
- Tata Communications,
- Facebook,
- PureLifi,
- RuralConnect,
- GSMA Intelligence, and
- Several smaller institutions,

all aiming to provide possible solutions for internet connectivity that is not based on traditional infrastructure and protocols, and could effectively be used for last mile internet connections in rural areas and in emerging markets where limitations and challenges stifle traditional solutions.

Each of these solutions are briefly reviewed in terms of the technology(ies) applied, its feasibility, and the backing and sustainability of each project. Of course the larger companies such as Google and Facebook have little financial limitations and should ideally be seen through towards commercially available products, but many smaller companies are also providing solutions very specific to certain areas in developing countries, with technology catered to the limitations of these areas.

6.4.1 Project Loon (Google)

Project Loon from Google, launched in 2013, is a balloon-based system that delivers connectivity from an altitude of 20 km up in the stratosphere, and aims to expand connectivity to unserved and underserved areas around the world. These balloons

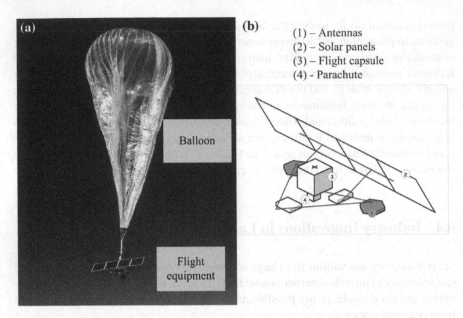

(b)
(1) – Antennas
(2) – Solar panels
(3) – Flight capsule
(4) - Parachute

Fig. 6.2 Google's Project Loon, showing the Loon system's **a** balloon and flight equipment, and **b** the primary equipment onboard the system. Figures obtained from official Project Loon website, at http://loon.com

are designed to endure harsh environmental conditions, typically associated with the altitude at which it operates, where winds upwards of 100 km/h and temperatures of − 90 °C are common. The system mainly comprises of a balloon, and flight equipment, obtained from the official Project Loon website at http://loon.com, and presented in Fig. 6.2a and b.

The balloon in Fig. 6.2a is manufactured from sheets of polyethylene. It can last for approximately 100 days before having to land back on earth. The flight equipment in Fig. 6.2a and b comprises of energy efficient systems, powered by renewable energy through solar panels, and consisting of batteries for nighttime operations. The primary antennas transmit connectivity from ground stations, across a mesh network of Loon balloons, and back down to serve users with an LTE connection. The balloons are autonomously navigated to reach countries around the world where connectivity is required. It transmits an operator's signal from connection points on the ground, autonomously distributes the signal through the Loon mesh network of balloons, and send the LTE signal back to a user on the ground.

To date, since its inception in 2013, and apart from the technological achievements that Project Loon has accomplished, the system has achieved several accomplishments in terms of providing connectivity in areas where little infrastructure exists. These accomplishments include

- connecting a school in the rural outskirts of Campo Maior, Brazil, to the internet for the first time,

- providing emergency connectivity for people affected by floods in Peru in 2017, and
- assisting users in Puerto Rico after Hurricane Maria in 2017, providing basic connectivity to over 200,000 people with balloon using machine learning algorithms to group and direct them to Puerto Rico.

Project Loon is a good example of technology capable of delivering last mile connections to users in remote and rural areas, and as the project matures, it could become a permanent fixture over such areas, providing users with a continuous and reliable internet connection.

6.4.2 BRCK (Kenya)

A Kenyan company, BRCK, is bringing internet access to East Africa, and has become the largest public Wi-Fi provider in Sub-Saharan Africa after its recent acquisition of Surf, another internet service provider [15]. In 2013, BRCK was formed by some of the co-founders of Ushahidi and iHub (Nairobi), primarily to mitigate Kenya's infamously undependable grid. Combined, BRCK currently (at the time of writing) has over 2000 public Wi-Fi hotspots across Kenya, with close to half a million unique active users each month.

The offered service, Moja Free, has three distinctive characteristics as outlined on its official website (www.brck.com), being

- free internet connectivity for anyone with a Wi-Fi enabled device,
- an Android application for an improve Moja Free experience whenever a Moja Wi-Fi signal is found, and
- free content which includes movies, television shows, and books.

Connectivity to the Moja Wi-Fi network (at 2.4 GHz with 5 GHz options also available) is supplied by hardware manufactured by BRCK, which is essentially a micro-computer with local hard drive storage space, redundant LTE, HSPA+, and GPRS failover connectivity options, an on-board battery if a power outage occurs, and sealed off in a weather resistant enclosure.

As an educational business model, BRCK Education provides kits of hardware, software, and connectivity tools to deploy smart classrooms in rural areas, already reaching learners in over 17 countries, providing digital education for students, a critical skill required for Industry 4.0 participation.

6.4.3 Tata Communications in South Africa

Tata Communications [19] presents a white paper on addressing the opportunities and challenges facing service providers and consumers in emerging markets. Tata

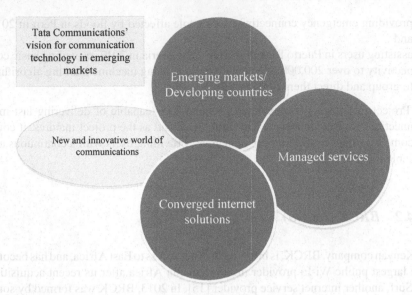

Fig. 6.3 Tata Communications's vision for communication in emerging markets—*a new world of communications*

Communications defines itself as being ideally positioned in a number of emerging markets around the world to act as a multifaceted service provider to the traditional service providers. Tata Communications aim to assist local service providers to deliver high-quality services to the end-users. In South Africa and India (also extended to Brazil in 2017), Tata Communications has a strong presence in last mile connectivity. The group has acquired numerous data centers, with a skilled workforce capable of running operations in these countries. Such capabilities also give them the opportunity to expand to nearby markets such as Pakistan, Sri Lanka, and Bangladesh. The vision of Tata Communications is outlined in Fig. 6.3.

As shown in Fig. 6.3, the vision of Tata Communications is to deliver high-quality and cost-effective communications to advance the reach and leadership of consumers, specifically in emerging markets such as India, Asia (including China), and in South Africa. This is achieved through converged internet solutions and managed services, such as cloud computing.

In India, Tata Communications owns and operates nationwide last mile access, with interconnection arrangements with the community of telecommunications service providers of India. As a result, they are able to provide high quality, efficient, accountable end-to-end internet capabilities, operated by a growing skilled workforce. According to Tata Communications, in India they operate

- 40,000 km of transmission network, spanning over,
- 300 cities and towns, with over,
- 200 points of presence (PoPs),
- 4000 km of metro fiber, and

- 5000 connected buildings.

Also according to Tata Communications, the primary advantages of this operation in India include:

- a global fiber backbone allowing access to global services and provide high levels of security and accountability of business-critical applications and data,
- a variety of last mile access capabilities through fiber, Ethernet, and WiMax,
- sourcing local equipment to create jobs within India and support local manufacturing, and
- using India as a base to deliver services to emerging markets throughout Southeast Asia.

In South Africa, Tata Communications owns and operates one of the largest subsea-level cable capabilities worldwide, and are the majority stakeholder in last mile access national carrier, Liquid Telecom. Their capabilities are further expanded on through a consortium of members of the SAT-3/SAFE and WACS cable systems. Interconnectivity with Europe, India, and Asia is established through the SEACOM system, launched by Tata Communications and Neotel in 2009. From the Tata Communications white paper, the three primary advantages of being a last mile service provider in South Africa are:

- A direct reach to the last mile and the user base, primarily through more than 10,000 km of national optical fiber, FTTx in major metropolitan business districts, as well as WiMax connections in four major cities in South Africa. At the core of this system, is an IP/MPLS network that offers quality of service reaching all of the major cities in South Africa.
- Abilities to support business-critical applications and services through a variety of redundancy implementations. Much of this redundancy is achieved through its major capital investment in the SEACOM system, serving the east coast of Africa. Tata Communications also invests in redundancy facilities on the west coast of Africa, and this fiber ring around the continent enables them to provide last mile access to all countries in Africa.
- Having data centers in Johannesburg and Cape Town, South Africa, Tata Communications is able to provide connectivity to business-enabling cloud-based services, where Neotel supports distinctive and locally sustained proficiency to reel out services that are integrated with Tata Communications' international strategy.

Much of Tata Communications's success stems from long-term relationships and partnerships with local telecommunication operators in emerging markets. Acting as an intermediary, they are able to facilitate relationships between service providers that have no pre-existing associations with local operators.

6.4.4   Facebook (A Link Between Electrification and Connectivity)

Research from Facebook and the University of Massachusetts Amherst shows that access to electricity in households have a strong causal effect on connectivity [2]. In this research, it shows that households recently connected to the electricity grid show a significant and lasting increase in device utilization and new account creation on the Facebook platform. Key results show that with each new grid connection,

- one additional smartphone connected to the platform, and
- 0.52 new users were actively engaging on the platform within the first 12 months [2].

According to the report, this is the first time that a causal relationship between better energy access and connectivity could be statistically shown. Energy access companies are using solar and battery storage to power connectivity in rural areas, with various business models, from small portable solar kits for individual households, to village-scale micro-grids that can power local businesses and cellular infrastructure. Alternative methods include phone-charging kiosks for individuals, grid extensions in households, and pay-as-you-go solar, also for households.

Expanding a cellular network typically involves the construction of new towers for the base stations that connect mobile phones to the wider network. However, remote rural areas have much lower customer density and has increased operating cost. Over a million cellular towers in developing countries are off-grid or have at best unreliable grid supply [2]. These towers rely on diesel generators for its primary power during large parts of the day to avoid interruptions to the mobile network during congested times. Although various methods exist to innovate and collaborate to bring connectivity to more people in emerging markets, including policies on

- distribution and logistics sharing among mobile network operations,
- sharing consumer data leading to better analytics and ultimately customer acquisition,
- alternative payment methods and using mobile money,
- deployment partnerships to eliminate struggles in sourcing reliable low-cost power at the last mile, and
- industry coalitions to accelerate innovation, build consumer awareness, and push for appropriate regulations.

As with updated policies, regulations, and collaborations, technology remains at the forefront of high-bandwidth connectivity for last mile connections, and small-cell networks and alternative infrastructure in rural parts plays a significant role according to Bloomberg [2]. A number of companies have been working on building such connectivity infrastructure, and to assist in accelerating progress, Facebook launched the OpenCellular project under TIP to develop open source hardware and software for small-cell base stations. One of the group members, Nuran Wireless, is aiming to complete a tower solution costing US $7000 with a 2G base station having a 1–2 km range. Power requirements for this tower is below 15 W, and the entire tower, which includes backhaul, operating at energy levels below 50 W. A similar project, a tower offering 4G LTE connectivity and operating below 55 W is also undertaken by Nuran Wireless. Each site varies as a function of population dispersion, revenue, and terrain, and the flexibility of these towers enables operators to optimize their deployments. Using modeled parameters from India shows that small-cell infrastructure can be more economical than diesel-powered macro towers situated less than 10 km outside of major cities.

In parallel, research into distributed rural access networks using unlicensed spectrum, such as Wi-Fi and TVWS have also begun to energy. Initiatives such as Express Wi-Fi, BRCK, and Mawingu are in the process of testing configurations to deliver broadband services to both urban and rural areas. These architectures have similar (low) power requirements when compared to small-cell towers, and is able to provide additional opportunities for companies that are able to supply the appropriate equipment and services, which includes on-site training of local skilled workers. As part of the OpenCellular project, Facebook also launched the OpenCellular Power, and open source project to support an ecosystem for power equipment specifically designed with the above-mentioned connectivity needs in mind.

6.4.5 PureLifi

PureLifi, arguably the champion and founder of commercially available Li-Fi (also working with the IEEE to form the global standard on Li-Fi, the 802.11bb standard), published [12] to discuss Li-Fi's relevance in Industry 4.0. It looks at some of the impacts that wireless communications in general will have on Industry 4.0 in manufacturing. The publication proposes that manufacturers take full advantage of advancements made in wireless technology and that smart factories would require extremely low *latency* wireless connections to facilitate high production throughput. Furthermore, the alternative of deploying Li-Fi, as opposed to radio frequency (RF)-based wireless technologies such as Wi-Fi, is beneficial in RF-sensitive areas such as oil, gas, chemical, nuclear, and healthcare facilities. PureLifi also debates that RF leaks (wireless signals that are susceptible to interception) are posing *security* risks—an inherent characteristic that Li-Fi is resistant to due to its line of sight requirements.

Therefore, PureLifi [12] highlights the two primary considerations that make Li-Fi the superior alternative in smart manufacturing, being

- low latency and
- better security.

According to PureLifi [12], in the manufacturing environment, various components such as programmable logic controllers, human machine interfaces, sensors, and motor drives require communications with low latency. In terms of RF, there are three primary limitations in Industry 4.0 manufacturing implementations, being its reflective nature leading to self-interference, electromagnetic interference, and RF leakage that can be hacked or jammed. PureLifi [12] performed experiments in latency and packet loss, comparing Wi-Fi and Li-Fi, and the results are summarized in Table 6.3.

As shown in Table 6.3, the latency and packet loss of Li-Fi is significantly better than that of Wi-Fi, as argued by PureLifi [12]. From PureLifi's official website (pure-lifi.com), experimental setups and projects have been deployed in over 20 countries to date, and numerous case studies are presented on this site. The primary categories where successful deployments are made are in

- the healthcare sector,
- engineering services and maintenance,
- training programs in the telecommunications industry,
- disaster operations,
- military ICT operations,
- performance lighting, and
- secure wireless connections in architectural firms, property developer firms, and tenants.

From PureLifi's official website, no reference to case studies on last mile projects in emerging markets could be found, at the time of writing, and several other websites were accessed to determine if such pilot programs or projects exist. Table 6.4 lists some of the websites access in January 2019.

Table 6.3 PureLifi [12] experiment on latency and packet loss of Wi-Fi and Li-Fi

	Wi-Fi	Li-Fi
Latency (μs)	2423	740
Packet loss (%)	35.6	1.8

Experiments were performed with UDP packets of 100 bytes sent between personal computers at varying intervals of 500, 1000, 1500 μs

Table 6.4 A list of major manufacturers and development companies in the Li-Fi space, provided with links to official websites and notes taken in January 2019 based on visiting these links

Company	Official website	Notes
LVX system	www.lvx-system.com	No products listed on site. Last update to site in 2016
Axrtek	www.axrtek.com	Homepage works—all other URL's not found
Velmenni	www.velmenni.com	Actively conducting pilot projects and research on optimizing Li-Fi
Oledcomm	www.oledcomm.net	Active Li-Fi technology patent and product portfolio
PureLifi	www.purelifi.com	Active Li-Fi technology patent and product portfolio. Partner in developing global IEEE standard

From the list provided in Table 6.4, it was found that only PureLifi, Velmenni, and Oledcomm were currently (January 2019) active in technology patents and research in Li-Fi technology. Again, no references were found of last mile solutions in emerging markets or rural areas.

Another industry innovation in last mile solutions with relevance in emerging markets as well as rural areas (due to its large coverage area), is television white space (TVWS), briefly reviewed in terms of its relevance in Industry 4.0 in the following paragraph.

6.4.6 TV White Space (TVWS)

White space describes vacant terrestrial broadcasting frequencies within the wireless band. In particular, television networks have large openings among channels used for buffering, and if not actively used or needed, this spectrum can be allocated to distribute pervasive high bandwidth internet. TVWS signals can travel up to 10 km as a function of its frequency, through foliage, large structures such as buildings, and further obstructions, as a function of its wavelength, described in Chap. 3 of this book. It therefore does not require LoS, and portable end-user devices (using additional hardware capable of decoding the signals) can access the WS through fixed or portable power stations. The amount of bandwidth that can be allocated to internet broadcasting varies with the instantaneous requirements of the TV broadcasting, but typical spectrum ranges between 470 and 790 MHz [11].

Companies such as Google and Microsoft have already explored TVWS in emerging markets, particularly in Africa, taking advantage of the already available spectrum and the large distance that these signals are able to traverse without requiring regular signal boosting. According to Dr. Apurva N. Mody [6], the chair of the WhiteSpace Alliance, an organization that promotes the deployment of WS for broadband internet, rural areas are the ideal regions to start expending WS technology. According to Dr. Mody, one of the largest issues in rural markets is to join the device points, since housing is generally separated by large distances, a challenge for fiber optics or traditional copper cables.

In 2013, Google assisted the city of Cape Town in South Africa, along with the local Independent Communications Authority of South Africa (ICASA) and the Council for Scientific and Industrial Research (CSIR), to launch TVWS trials to provide wireless broadband backhaul to 10 schools in the Western Cape, South Africa. The trial, named RuralConnect and designed by Carlson Wireless Technologies, was designed to explore whether TVWS is a viable solution for providing broadband to more people in South Africa, especially in rural communities. The RuralConnect TVWS scenario is depicted in Fig. 6.4.

The RuralConnect trial was designed, as shown in Fig. 6.4, to deliver high-bandwidth and low-latency connectivity using an access point and a high-gain omni-directional antenna, ideally powered with renewable energy such as solar cells. With a 10–40 km operating radius, the wireless signal is then broadcast to a specifically

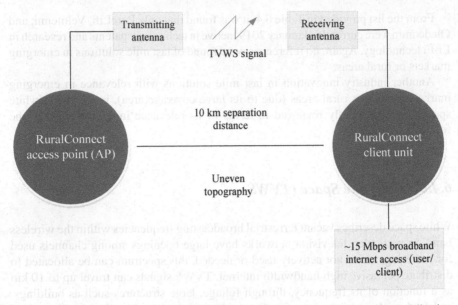

Fig. 6.4 A typical scenario depicting TVWS used for wireless broadband backhaul, delivering internet connectivity through unused spectrum, adapted from Gilpin [6]

designed RuralConnect client unit. At the end of the trial period (September 2013), it was up to the local authority, ICASA, to guide regulation of TWVS. As of April 2018, ICASA published the final regulations of the use of TVWS, although deployment of the technology is still unknown [10]. From the operational statistics obtained through the 6-month trial in South Africa, the TVWS broadband throughput (for downloads) ranged between 2 Mbps (during peak periods) and 12 Mbps (the maximum uncontested throughput). Upload throughput ranged between 1.5 Mbps and approximately 4 Mbps, throughout the 6-month period [20].

Microsoft's 4Afrika initiative (launched in 2013) also uses WS technology throughout the African continent, with projects running in South Africa and Tanzania, aimed to bring internet connectivity to the millions of African individuals that do not have internet access. The initiative invests in small-to-medium enterprises (SMEs), local governments, and the youth in Africa, focusing on delivering affordable internet access, increasing trained personnel, and financing native technology resolutions. Since 2013, 4Afrika has extended to over 1.7 million SMEs, getting over 500,000 of these enterprises connected to extent new markets and cultivate their respective industries, where TVWS connectivity pilot programs running in last mile communities have played a large part in bringing internet connectivity to schools, healthcare centers, universities, and businesses.

In South Africa, a startup company named AfriCanopy, aims to solve the problem of high prices of mobile data, leading to people living in rural areas unable to access the internet, through TVWS broadband internet access [9]. AfriCanopy is set to provide 85,000 residents and 50 schools with low-cost internet access, while creating

400 jobs in the process (South Africa has an unemployment rate of 27.5% as of the third quarter of 2018 [21]). Base stations implemented on this technology are 16 km apart, and provide coverage of up to 20 km, provided there is line of sight to connected devices. To mitigate outages due to non-line of sight connections, Wi-Fi mesh networks are constructed to extend the range beyond obstructions. Each base station is capable of aggregating 20 MHz of bandwidth, with peak capacity of 169 Mbps. Its pricing structure would allow for single megabyte purchases at prices that are comparable to the current pricing structure from local telecommunication providers if data is bought in bulk, for example a 20 GB purchase.

6.4.7   GSMA Intelligence

GSMA Intelligence [8] acknowledges that although there is a significant growth in pervasive computing through mobile networks worldwide, a large portion of the global population still suffers from minimal coverage, unable to take advantage of the benefits of ubiquitous connectivity. The final frontier (last mile) has become one of the most overtly noticeable platforms for a multitude of trials with unconventional methods of connectivity, backed and promoted by big internet players, with wireless connectivity and the usage of WS in the television frequency band (TVWS—briefly reviewed in the following paragraphs) being of the most prominent. GSMA Intelligence realizes that these alternative technologies are pursuing prospective use cases to grow connectivity past current mobile infrastructures and assist in driving socioeconomic influence in developing countries, particularly through expanding internet access to isolated rural districts and disaster reaction regions.

According to GSMA Intelligence [8], connectivity is expanded through balloons (Google's Project Loon), drones (Facebook), and satellites, all benefiting from a broader array of ground coverage, especially useful in aiding secluded rural populations and in certain cases facilitating backhaul capacity. These aerial networks are inherently intended to maximize ground coverage through the benefit of elevation. In these scenarios, a consumer is able to join the network by transferring signals from an end-user device upwards to a localized balloon, drone, or satellite, operating through a mesh network that is also able to connect with ground stations and finally back to the consumer. Figure 6.5 is an indication of the altitudes of aerial networks as a function of ground coverage.

From Fig. 6.5, as the networks become unexposed of terrestrial impediments (nLoS), a niche-use case is presented for connecting people in secluded rural districts that fall out of the range of terrestrial mobile signals. Important to note is that the mobile signal quality weakens and decreases as a function of altitude (distance), as shown in Chap. 3 of this book. GSMA Intelligence highlights that although these airborne networks implement high-frequency spectrum allocation, in concept, it could facilitate both voice and data links. It however appears [8] that the priority is placed on data traffic, given inherent risks of long distance electromagnetic signal latency and bandwidth requirements for practical applications (for example

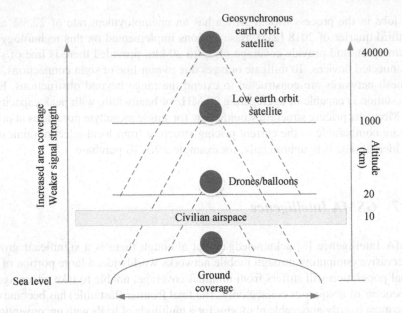

Fig. 6.5 A representation of the altitude of aerial networks and the coverage gain as a function of altitude, primarily for balloons, drones, and satellites

in disaster response situations where latency and unreliable connections are often avoided). Ideally, these aerial networks can connect to localized base stations on the ground, capable of demanding signal processing and serving as last mile connectivity to users, through either wired (fiber or Ethernet) connections, or a wireless backhaul, ideally in the mm-wave spectrum.

6.4.8 CSquared

Initially named Project Link, renamed in 2013 to CSquared under a new Google brand, in 2011 this project identified a deficiency of fiber optic interconnects in big African capitals as a major barrier to affordable and reliable internet access [7]. CSquared has constructed over 800 km of optical interconnects in the metropolises of Kampala and Entebbe, and 840 km fiber in Accra, Tema, and Kumasi in Ghana. Over 25 internet service providers are making use of these metro fiber networks in both Ghana and Uganda, offering broadband services and wireless 4G solutions to end-users, achieved through the over 1200 tower and commercial building locations connected to the CSquared fiber network. Through the Research and Education Network for Uganda, CSquared's fiber scheme delivers broadband last mile connections for tertiary education and establishments conducting research within the health sector situated in the Greater Kampala Metropolitan Area. Furthermore, CSquared incorporates partnerships with locally operated institutions to develop a technology sector invested in Sub-Saharan Africa, also contributing to developing skilled workers and entrepreneurs in these communities.

6.4.9 Winch Energy and iWayAfrica

Winch Energy and iWayAfrica operates a pilot program that aims to install a remote *solar* kiosk (the Winch Hub), with incorporated VSAT apparatus that provides broadband connectivity. The project will be made accessible to five groups in Bunjako Island, a fishing community on the coasts of Lake Victoria, and Uganda, reaching a populace of approximately 20,000 people. This Winch Hub is therefore an off-grid solution, specifically aimed at providing connectivity to suburban, civic, and business consumers in rural settings. The African country Uganda was specifically chosen since the country has less than 20% reliable means of electricity, and approximately 65% of the country does not have access to internet [4].

6.5 Conclusion

The lack of high quality last mile solutions in emerging markets and rural areas is preventing societies from taking advantage of the benefits of information sharing, it is important to reflect on implementations, programs, and initiatives that spearhead last mile connectivity in both developed countries and in emerging markets. The implementations in developed countries are often without (or at least with very little) financial constraints, a combination of the newest matured technologies and high-quality components, maintained by skilled workers, and sustained/upgraded through its lifetime. These solutions have a significant importance when developing last mile solutions in developing countries, specifically in terms of the log-term sustainability of these solutions.

In emerging markets, however, the additional limitations as discussed in Chap. 2 of this book, mean that implementations of last mile connectivity are inherently different, albeit based on solutions in developing countries. In a rural dwelling, in a developing country, with minimal access to internet, skilled workers, and even electricity, a solution to empower the local citizens through information is vastly different to that of many other areas. Poor areas also have limited access to technology that is capable of rendering internet content, such as smartphones, tablets, or traditional computers. Therefore, exotic solutions of internet distribution are not practical, and solutions are needed that make use of mature technologies such as Wi-Fi or GSM. This book also argues for technologies such as Li-Fi, which would require certain infrastructure considerations to roll out. An advantage of lack of infrastructure (or very little infrastructure) in such rural areas, is that new and novel technologies such as Li-Fi can be used with minimal overheads to infrastructure development. Furthermore, combinations of technologies to serve local users provide mature and modular solutions in specific scenarios. Classrooms, local clinics and other medical practices, and office space can implement Li-Fi solutions for desktop and laptop computers and other machinery, whereas Wi-Fi can provide a more mobile solution within the area, both using fiber or mm-wave backhaul, and (satellite or balloon-based) GSM can be used when travelling outside the coverage area of Li-Fi and Wi-Fi.

In this concluding chapter of *Last mile internet access for emerging economies*, the opportunities in emerging markets are discussed in terms of the potential to provide socioeconomic benefits to citizens keen and eager to expand their knowledge, client-base, and entrepreneurial opportunities through the internet. Many people do not have access to the internet, this is a well-known fact, and this chapter also briefly reviews the statistics of the last billion, a term for the poorest billion citizens globally—with specific reference to the resources available to these citizens. The section also refers to the severest challenges faced by the last billion, and reviews a global change in mindset required to provide this group of people with necessities that have the potential to provide socioeconomic growth and some sort of sustainability. It is realized that internet access might not be the top priority for the last billion, but it is argued that over the long-term it would become essential, particularly in providing opportunities for these citizens to participate in the global economy, especially since the fourth industrial revolution would provide such opportunities, given internet access.

To show that numerous corporations share the ideal of providing internet access to as many people on earth as possible, various industry innovations are presented in this chapter. These innovations, albeit some still in prototyping phases, or rolled out to smaller control populations, it still showcases technologies that are capable of providing effective last mile solutions in areas where traditional connections are difficult to achieve. Large corporations such as Google and Facebook, as well as Tata Communications are spearheading such movements, but many smaller companies such as RuralConnect, Winch Energy and iWayAfrica are also providing the necessary infrastructure research and developments needed to find sustainable last mile solutions. Li-Fi developments are also reviewed in this chapter, however, no last mile solutions for emerging markets and rural areas were found, possible showing a significant opportunity for potential Li-Fi service providers.

Finally, this book discussed and reviews not only the challenges and limitations that emerging countries are faced with, but also offers potential solutions through a technological review of technologies capable of providing tailored last mile solutions. The first chapter of this book briefly looks at the alternatives available that can achieve this, whereas Chap. 2 describes the specific challenges and limitations in developing countries with respect to internet connectivity. Chapter 3 provides a review and overview of signal propagation and networking, the knowledge required to effectively determine the ideal solutions in a variety of environments. Chapters 4 and 5 reviews last mile technologies that are not as mature as for example Wi-Fi and GSM, but offer numerous advantages especially in distributing internet in areas with very little infrastructure.

References

1. Becker C (2017) Emerging markets, emerging technologies. Retrieved 11 Jan 2019 from http://www.investec.com
2. Bloomberg (2018) Powering last-mile connectivity. Retrieved 15 Nov 2018 from http://data.bloomberglp.com
3. Boumphrey S, Verikaite R (2017) Six tips for working with emerging market data. Retrieved 11 Jan 2019 from http://blog.euromonitor.com
4. Crown Publications (2018) Bringing energy and internet access to rural communities. Retrieved 19 Jan 2019 from http://crown.co.za
5. Friedrich R, Ward J, Singh M, Lesser A (2007) The mobile broadband opportunity in emerging markets. Retrieved 11 Nov 2018 from http://www.4gafrica.co.za
6. Gilpin L (2014) White space, the next internet disruption: 10 things to know. Retrieved 11 Nov 2018 from http://www.techrepublic.com
7. Google Blog (2017) CSquared gets new investors to expand internet access in Africa. Retrieved 17 Nov 2018 from http://africa.googleblog.com
8. GSMA Intelligence (2014) Analysis. Mobile access—the last mile. Retrieved 11 Nov 2018 from http://www.gsmaintelligence.com
9. McKane J (2018) The new mobile network for rural South Africa—how it works. Retrieved 20 Jan 2019 from http://mybroadband.co.za
10. Mzekandaba S (2018) ICASA makes headway with TV white space regulations. Retrieved 12 Nov 2018 from http://www.itweb.co.za
11. Nyasulu T, Anderson D, Crawford DH, Stewart RW, Brew M (2015) TV white space for internet access in the developing world. Centre for White Space Communications, Dept. Electronic and Electrical Engineering, University of Strathclyde, Glasgow
12. PureLifi (2016) LiFi making industry 4.0 a reality. Retrieved 27 Jan 2019 from http://cdn2.hubspot.net
13. Qiang C, Rossotto C (2009) Economic impacts of broadband. Information and communications development
14. Schmida S, Williams I, Lovegrove C (2016) Business models for the last billion: market approaches to increasing internet connectivity. Agency for international development of the United States government.
15. Shapshak T (2019) Kenya's BRCK acquires surf to become the biggest public WiFi network in Sub-Saharan Africa. Retrieved 15 Feb 2019 from http://www.forbes.com
16. Shengling B, Simonelli F, Bosc R, Zhang R, Li W (2017) Digital infrastructure: overcoming digital divide in emerging economies. G20 Insights, 3 Apr 2017
17. Statista (2019) Global information and communication technology (ICT) revenue from 2005 to 2019 (in billion euros). Retrieved 13 Jan 2019 from http://www.statista.com
18. Stats SA (2018) Post and telecommunications industry, 2016. Report no. 75-01-01. Embargoed until 27 Sept 2018
19. Tata (2017) Tata communications extends emerging market reach to Brazil. Retrieved 18 Jan 2019 from http://www.tatacommunications.com
20. TENET (2018) The Cape Town TV white spaces trial. Retrieved 12 Nov 2018 from http://www.tenet.ac.za/tvws
21. Trading Economics (2019) South Africa unemployment rate. Retrieved 20 Jan 2019 from http://tradingeconomics.com

Printed in the United States
By Bookmasters